INDUSTRIAL ENGINEERING MANUAL
for the Textile Industry

SECOND EDITION
REVISED AND ENLARGED

INDUSTRIAL ENGINEERING MANUAL

for the Textile Industry

Edited for
ASSOCIATION OF TEXTILE INDUSTRIAL ENGINEERS

by NORBERT LLOYD ENRICK

Professor of Administrative Sciences, Kent State University, Kent, Ohio

Formerly Associate Director and Head of Operations Research Division, Institute of Textile Technology, Charlottesville, Virginia

ROBERT E. KRIEGER PUBLISHING COMPANY
HUNTINGTON, NEW YORK
1978

Original Edition 1962

Second Edition 1978, revised and enlarged

Printed and Published by
ROBERT E. KRIEGER PUBLISHING CO., INC.
645 NEW YORK AVENUE
HUNTINGTON, NEW YORK 11743

Printed in the United States of America

Library of Congress Cataloging in Publication Data

Enrick, Norbert Lloyd, 1920– ed.
Industrial engineering manual for the textile industry.

Bibliography: p.
Includes index.
1. Textile industry–Management. 2. Industrial engineering. I. Title.
[TS1445.E49 1978] 658'.97'7 77-15461
ISBN 0-88275-631-1

Preface to Second Edition

This manual, similar to the earlier *Time Study Manual for the Textile Industry,* represents the work of a large number of contributors affiliated principally with the Southern Textile Methods and Standards Association. It is from this wealth of practical experience, know-how, study, and research of these individuals that the present manual was developed.

To make this a comprehensive and complete work of widely applicable value in industrial engineering fields, a broad view of the topic given in the title has been taken. As a result, there are included several chapters on the modern techniques of operations research, with emphasis on the types of applications that are particularly useful to the industrial engineer. Numerous case histories illustrate practical applications. Finally it must be recognized that industrial engineering is increasingly an activity requiring interlocking considerations of broad economic, financial and marketing factors. The relevant data from which an industrial engineering program consistent with all major functions of the organization must be developed and operated are comprised in the firm's management information system (MIS). In the final chapters of this book, therefore, the principal features of installation, operation and control of MIS at various levels of usage are discussed.

Whereas the general principles and methods of industrial engineering and operations research are described in several texts, there has been a need for a special manual that considers in its scope the special problems of continuous and semi continuous processing and the complex operating problems of textile mills and mill organizations. This Manual has been designed to fill these needs and also to serve as a study, reference and training aid.

Additionally, this book should be of interest to managerial, technical and supervisory personnel who wish to gain a fuller knowledge of industrial engineering and the related operations research and MIS applications as a cost-saving and profit-building factor in effective mill operations.

A prior reading of the *Time Study Manual* is not needed. The present volume is self-contained.

Kent, Ohio
June 1978

Norbet Lloyd Enrick
Editor

Preface

This manual, similar to the earlier *Time Study Manual for the Textile Industry*, represents the work of a large number of contributors affiliated principally with the Southern Textile Methods and Standards Association. It is from the wealth of practical experience, know-how, study, and research of these individuals that the present manual was developed. The original material for this work consisted of close to a dozen papers on various time study and industrial engineering subjects.

To make this a comprehensive and complete work that would be of practical value in industrial engineering fields, the editor obtained additional material from members of the Association and their colleagues. Included are several chapters on the modern techniques of operations research—an area which is considered especially pertinent to applications that can be made readily by the industrial engineer. Appropriately, the treatment of this important subject is accompanied by numerous case histories.

Whereas the general principles and basic methods of industrial engineering and operations research are described in a number of good texts, there has been a need for a special manual that considers in its scope the problems of continuous and semi-continuous processing and the complex operating problems of the textile mill. *Industrial Engineering Manual for the Textile Industry* has been designed to fill this need and also to serve as a study, reference, and training aid for the industrial engineer.

The book should also be of interest to management, and technical and supervisory personnel who wish to gain a fuller knowledge of industrial engineering as a cost-saving and profit-building function in modern mill management.

Although desirable, a prior reading of the *Time Study Manual* is not essential, as the present volume is self-contained. The material is understood readily and may be applied without a previous knowledge of the subject matter.

Charlottesville, Virginia
January, 1962

NORBERT LLOYD ENRICK
Editor

Contributors

ERNEST F. BALFREY, *Collins and Aikman Corp., Albermarle, N. C.*

ROBERT G. CARSON, JR., *School of Engineering, North Carolina State College, Raleigh, N. C.*

JAMES M. COOPER, *Firestone Textiles, Gastonia, N. C.*

NORBERT L. ENRICK, *Associate Professor, University of Virginia Graduate School of Business Administration, Charlottesville, Va.* (formerly *Associate Director of Research and Head of Operations Research Division, Institute of Textile Technology, Charlottesville, Va.*)

BYRON GRIFFIN, *West Point Mfg. Co., West Point, Ga.*

ROBERT E. HEILAND, *Research Director, Kurt Salmon Associates, Inc., Coopersburg, Pa.*

DAVID B. HERTZ, *Principal in Charge of Operations Research, Arthur Andersen & Co., New York, N. Y.*

CHARLES M. JORGESON, *B. F. Goodrich Textile Products, Thomaston, Ga.*

CHARLES L. LAND, *Frank Ix & Sons, Lexington, N. C.*

SUE C. LAWRENCE, *Consultant, Williamsburg, Va.* (formerly *Research Associate, Institute of Textile Technology, Charlottesville, Va.*)

HOWARD L. LOVELESS, *School of Engineering, University of Tennessee, Knoxville, Tenn.*

RICHARD E. MACHAN, *Textile Division, Goodyear Tire & Rubber Co., Cedartown, Ga.*

RICHARD W. MALONE, *Riegel Textile Corp., Ware Shoals, S. C.*

JAMES D. MILLER, JR., *Crown Cotton Mills, Dalton, Ga.*

THOMAS F. O'CONNOR, *Consultant, Oakland, Calif., and Charlottesville, Va.*

WALLACE J. RICHARDSON, *Department of Industrial Engineering, Lehigh University, Bethlehem, Pa.*

ROBERT D. SELLERS, JR., *Southern Bleachery & Print Works, Inc., Taylors, S. C.*

KURT H. SCHAFFIR, *Manager of Operations Research, Arthur Andersen & Co., New York, N. Y.*

W. S. SMITH, *West Point Mfg. Co., West Point, Ga.*

Contents

CHAPTER 1

Introduction

The past decade has witnessed a strong emphasis on the application of sound industrial engineering principles to all aspects of production in up-to-date and competitive textile mills. There is every evidence that these applications will increase many-fold in the decade to come. Matters that previously were considered too complex to submit to systematic analysis, planning, and control, are today being attacked with newly developed management tools. The resultant cost savings and improvements in quality and production have often been phenomenal.

This manual presents the most important of these techniques known today. The first five chapters are concerned with practical aspects of advanced time study, covering the application of measured standards, the maintenance of such standards, the determination of frequencies of elemental occurrences, and the handling of random occurrences. Two further chapters present the modern and popular techniques of work sampling in the textile mill.

Subjects falling into the more general classifications of industrial engineering, including job evaluation, cost systems, materials handling, automation, and quality control, are presented next. Several years ago a book on industrial engineering might have ended with these or similar subjects; however, a new set of techniques, known as *operations research,* have since become available to the industrial engineer. Accordingly, the most important methods of operations research that have found textile mill applications are covered in Part III of the book.

Operations research may be defined as a scientific method, using statistical and econometric tools, for the analysis, comparison, and evaluation of various alternative arrangements of mill operation. This evaluation may proceed on the over-all, the departmental and machine operator, and the processing detail levels. From the evaluation of alternatives the best arrangement can then be selected. The use of statistical and econometric methods permits us to evaluate alternative methods, procedures, setups, or courses of action in quantitative terms, so that the process of arriving at the optimal decision is facilitated. However, personal evaluation and judgment based on experience are still very much required.

Operations research provides valuable quantitative information that is an important aid to the decision-making function of all levels of management. Those who might doubt the value of operations research are invited to take some of the case histories presented in this book and compare their evaluation of the optimum with that derived mathematically.

It should be pointed out that whereas the statistical and mathematical procedures in this book are founded upon scientifically sound principles (which can be rigorously proven), the actual procedures presented have been purposely demonstrated so as to be readily understood by anyone familiar only with simple arithmetic. In place of mathematical proofs, intuitive explanations and demonstrations are offered. References to more detailed material are made where this is appropriate.

The particular operations research techniques discussed are queuing theory and related applications in the evaluation of machine interference in multimachine assignments. The important subject of *linear programming* is discussed with the aid of two actual, but slightly simplified, case histories. Evolutionary operation is described by using a slashing-weaving efficiency application. In conclusion, operations research techniques in the optimization and control of inventories are illustrated.

Industrial engineering works most effectively when supported by a well-designed system of automated data processing. Based on an understanding of the essential functions of a management information system (MIS), such as covered in the final chapters of this book, the industrial engineer will be able to participate in the design, operation and use of a system that includes his specific data needs.

PART I

WORK MEASUREMENT

CHAPTER 2

Application of Measured Standards

Introduction

The dictionary defines *measured* as "carefully considered" and *standard* as "a weight or measure to which others must conform." Therefore, a *measured standard* is "a carefully considered measure to which others must conform."

Standards must be applied to the productive equipment in such a manner that the worker will have a fair return for his efforts, and management will get the job done at a cost that will allow a fair return on the stockholders' investment. Assuming the standards to be in line, with due fatigue and personal allowances added for unusually heavy elements, the working time for all workers may be the same (60 minutes per hour less time for personal requirements). In some plants the allowances are given as definite work stoppages, such as two rest periods plus a lunch period per 8 hours, all paid at a basic hourly rate. In others, where it is not practical to shut off the power, the piece rates allow 10 to 15 per cent for personal and fatigue time, with the worker taking time off as needed.

Assuming that each operator can be assigned to about 52.50 minutes of work per hour (or 420 minutes per 8 hours), the problem then is to determine a piece rate by discovering how many machines will consume the operator's time and what rate of production may be expected. The piece rates in turn indicate the cost per pound for the various styles—an excellent means of cost control.

Production Constants

The industrial engineering department, being responsible for piece rates, may also be held responsible for the control of production. This involves the issuance of correct production constants to the payroll department for use in determining the total pounds or picks produced on each style by workers in the carding, spinning, spooling, winding, and weaving operations. By applying the production constants (such as pounds per hank) to the total hanks of each style produced, the total pounds are determined for each department, and should agree with those of the preceding and subsequent processes.

When paying by hanks, the production constant is simply the number of spindles per frame divided by the yarn number. This works very well on drawing and fly frames where contraction from twist is low. However, unless an adjustment for contraction is made on the spinning constants, the records will indicate a higher spinning production than the card room delivers, with a resultant standard cost understated by 6 to 10 per cent. In order to correct the spinning production constants and get the production in line with the card room, it is necessary to use the equivalent uncontracted counts over the rolls, rather than the contracted counts on the bobbin, as the divisor in the above calculations.

If the production between processes remains unbalanced it would be well to check the stampings on hank clocks, especially in the spinning and twisting processes, because it is very possible that the clocks were set up originally to produce long hanks to partially offset the yarn contraction.

Hank clocks on 1 $^1/_8$ inch roll fly frames may be stamped "1 $^1/_8$ inch, 19:45," indicating that the clock is set for 84 yards per tenth hank, or 840 yards per hank, since $19 \times 45 \times 1\ ^1/_8 \times 3.1416/36 = 84$ yards. Some clocks on 1 $^1/_4$ inch roll fly frames may be stamped "1 $^1/_4$ inch, 20:39," indicating 85 yards per tenth hank or 850 yards per hank. On spinning frames the stamped gear ratios may be "1 inch, 25:41," which equals about 89 yards per tenth hank or 890 yards per hank, since $25 \times 41 \times 1 \times 3.1416/36 = 89$ yards.

Twisters with 1 $^1/_2$ inch rolls may have hank clocks marked "1 $^1/_2$ inch, 890" or "1 $^1/_2$ inch, 20:34," each of which equals 890 yard hanks. With markings of "1 $^1/_2$ inch 840," "1 $^1/_2$ inch, 20:32," or "1:640," they are 840 yard clocks.

If 890 yard clocks are found it is necessary to adjust the production constant upward by the proportion of 890/840 or 1.06.

Establishing Standard Machine Efficiencies

Before equitable workloads can be calculated or assigned, it is necessary to determine the standard machine efficiencies. Whether or not proper assignments are in effect, it is necessary to make time studies of each job and to determine the length and frequency of its various elements, the machine interference, and the factors affecting the job. From such data, unit times are established. Hank clock readings taken before and after time studies should be recorded for use in confirming the calculated machine efficiencies expected from application of standard times and necessary waiting allowances.

Methods used to establish standard machine efficiencies vary for nearly each operation in textile manufacturing because of the differences in processing. Some machines, such as cards, continue to produce while creeling new laps and doffing full cans, losing production only while being stripped or ground. Drawing frames are stopped for creeling, doffing of cans, and for sliver breaks. Fly frames stop for doffing, creeling, some cleaning operations, and roving breaks. Spinning frames are stopped only for doffing, all other work being performed while the frames are running. Twisters stop for doffing and creeling, although those equipped with beam creels stop when a yarn breaks. Looms stop for warp and filling breaks, cloth roll doffs, beam run outs, and for repairs and resettings. Allowances for machine interference may be made on the basis of time study observations or calculated as shown in Chapter 15.

Each process should have a regular cleaning schedule, copies of which should be used as a guide during time studies. Often the cleaning frequency may vary for the different grades and types of stock processed.

Varying creel package and bobbin sizes will require different handling times for tenders and doffers, so that unit times for all types and sizes must be secured. End breaks per frame doff, or per 100 or 1000 active spindles per hour, must be determined for all styles. These data can be accumulated during observations.

Calculation of Workloads and Piece Rates for Card Tenders

Although it is possible to calculate card production from the doffer speed, some variations will be found owing to the thickness of the fillet, which must be added to the doffer diameter. Also, the tension draft involved between the doffer and calender roll must be considered. Therefore, the most practical point from which to calculate production is the calender roll. With use of a speed indicator or tachometer the revolutions per minute may be clocked and then converted to yardage or poundage of production.

After having determined the frequencies and applied unit times to them, it is a simple matter to calculate proper assignments, as well as the resultant labor savings to be expected from methods changes. For instance, if a tender has to transport his own full cans to the draw frames, and it will require about 3.5 minutes per card per day, then 25 cards will require 87.5 minutes per day. By assigning a can hauler, who is usually paid a lower hourly rate than a card tender, to do this work it is possible to reduce the labor cost and to increase the number of cards per tender to about 33, without increasing the workload.

Draw Frame Tenders

Production of the draw frames may be determined from the front roll speed, if the tension draft to the calender rolls is known. The calender rolls may be easily checked directly with a tachometer and these speeds used without further adjustments.

When calculating draw frame production efficiency it is necessary to allow for stoppages while the frame is waiting for service. The proper time allowance can be best determined by the methods shown in Chapter 15, supplemented by several 8 hour time studies. Where some of the elements are performed by the tender while the machine is either running or stopped, the time must be so noted. Thus, a time study may indicate the stoppage for creeling a frame to be 10.24 minutes, whereas the tender's total time for creeling is 13.12 minutes. In this case the tender prepared the cans from the cards by pulling out the pushed-in tops while the frames were running.

In order to calculate the frame efficiency without machine interferences, divide the running minutes per doff by the sum of the running time plus the stoppage time for doffing, creeling, repairing breaks, and cleaning per doff.

Where more than one drawing process is involved it is generally the practice to pay only for the hanks produced on the finisher process. On this basis, the production constant for pounds per frame hank is also applied only on the final process.

Slubber Tenders

Generally, the spindle speeds for slubbers will remain constant for each individual hank roving run. The production is governed by the front rolls, which in turn are controlled by twist gear changes. Checks may be made with a tachometer to determine the spindle speed. Next, roll speeds may be calculated from the turns per inch added. It is well to take periodic roll speed checks to assure use of proper twist gears.

When time studies are made on slubber tenders, it is necessary to record whether frames are running or stopped while work is performed, so as to accumulate the data needed in calculating frame efficiencies as well as tender assignments. The length of time studies depend on doffing, creeling, and cleaning frequencies, and generally a series of 8 hour studies are preferred, followed by 2, 4, and 5 hour studies to cover special elements (e.g., infrequent cleaning items and miscellaneous working times).

Frame efficiency, without machine interference, may be calculated by dividing the running minutes per doff by the sum of this value plus the stoppage time for doffing, creeling, repairing end breaks, and cleaning per doff.

On jobs where three or more frames are assigned to one tender, it is possible that machine interference will be high. It may then prove economical to assign a team of workers to a large set of frames. In charge of the set would be the tender at the normal piece rate, several creelers at a lower rate, and a helper at the minimum rate. Piece rates may be set up for each job category, with the poundage cost calculated from the total of the various rates per hank for each style.

Spinners

Since spinning frames are normally stopped only for doffing and periodic overhauls, it is comparatively simple to establish standard machine efficiencies. Where relatively few yarn numbers or bobbin sizes are involved in a set, doffing cycles can be planned to keep frame stoppages at a minimum.

Spinners' jobs consist primarily of creeling of roving, piecing of end breaks, cleaning, and patroling. Where practical, some of the cleaning should be assigned to others, such as roll picking, and other similar work above the ring rails. In smaller mills, where few frames are involved, it may prove more economical to assign all of this work to the spinner.

Although spinners may be paid on the basis of a rate per side hour, it is to the mill's advantage to pay by side hanks or frame hanks to avoid excessive frame down time.

Doffers

Sides or frames assigned per doffer may be calculated properly only after unit times for the various elements of the job have been determined. So many variables may be encountered that the elements should be broken down as much as possible, and tables be set up to cover the various combinations. For instance, the doffing time may include the following elements: secure doff boxes, stop frame, wind rack, doff frame, start frame, piece up ends, and deliver yarn. Because more boxes may be required per frame doff for large bobbins, the unit time should be higher than for smaller bobbins. The work to stop frames and wind racks may be uniform for all styles and bobbin sizes. The time required for doffing the frames will increase for the larger bobbins, as will the time for piecing ends after the doff. The delivery of the boxes of yarn after doffing may vary, depending on the boxes per doff and the delivery point.

The industrial engineer will be able to offer methods improvements when jobs of this type are properly broken down. Use of conveyor lines or tray elevators to the spooling or winding processes may be found to be more economical than manual labor. Overhead frame cleaners with air trunks may be a practical replacement for manual cleaning.

Where doffers are paid on the basis of side or frame doffs, there is a possibility that checks will be credited for doffs not made. Payment from hank clock readings tends to avoid this possibility, and also tends to keep the frames running. For cost purposes, pounds per side or frame doff may vary due to short doffs. However, pounds per hank is a constant figure for each yarn number.

Barber-Colman Spooler Tenders

Several methods are in use to pay spooler tenders. The simplest, but least desirable for cost control purposes, is to pay hourly rates. Other methods include side doffs from spinning, rate per 100 cheeses produced, or per trident of 12 or 15 cheeses, and rate per number of cycles of the knotting head per hour.

When a rate per hour is used, the same average cost per pound will prevail for all styles resulting from total payment divided by total pounds produced per day or week. If payment is on side doffs from spinning, short doffs are credited at the same cost as regular doffs, and it may be difficult to keep the production between processes balanced. Payment based on number of 100, or dozen cheeses, or tridents of cheeses, is more accurate for cost control purposes but requires constant attention to end break percentages, cheese yardages, and bobbin sizes. Most

Barber-Colman spoolers have a clock which records the number of cycles of the knotter head, so that a single rate per cycle can be used for all cases where the knotter cycle time is longer than the running minutes per bobbin. Retie-checks must be made to determine the new bobbins per cycle for production cost control purposes.

When time studies are made it will be noted that bobbins from larger rings on spinning frames may require increased handling time, although the time for cheeses doffed from all bobbins remains the same. Therefore, when setting up standard unit times for the various bobbin sizes, the items of creel, and retie per cycle may vary, whereas the other elements remain constant for all bobbins.

Winder Tenders

Yarn winding may be one of the most expensive single items in the total labor cost of a mill. However, by applying standard times determined from extensive time studies, it is possible to predetermine spindle assignments and pounds produced for various styles and package put-ups.

There are two basic types of winders: drum winders and precision winders. On drum winders, the package rests on a revolving drum and either a reciprocating thread guide or rotating traverse, cut in the drum, guides the yarn to the package. Precision winders have constant-speed spindles which turn the package, and yarn speed increases with increasing package diameter.

A simple and satisfactory method of calculating winder tender assignments is based on the standard working minutes per putup package, by using data on bobbins per package, reties per package, and gauge and doff time per package. Minutes available for winding divided by minutes required to produce one package yields packages per hour. Spindles per assignment are found by dividing the product of running minutes per package times packages per hour by the product of predetermined efficiency for each put-up, multiplied by 60 minutes per hour.

Both types of winders require the following steps: creel new bobbins, remove empty bobbins, retie end breaks, gauge and doff full packages, secure and tag empty packages, and start package winding. In general, the definite stoppages scheduled for blowing off and cleaning the winders must be considered along with personal allowance time when setting up piece rates. Rates may be based on side doffs from the spinning frames, dozens or other units of package produced, or 100 pounds produced and weighed before going to the conditioning room or before packing for shipment. If sufficient spindles are available, and runs of styles are large enough to warrant it, it may be more economical to wind from Barber-Colman spooling cheeses. For cost comparisons it should be considered that to the extra direct labor cost of spooling must be added the spooling department indirect labor and overhead charges. Another alternate that may be considered, in an effort to reduce winding costs, is the use of automatic winders such as the Abbott Traveling Spindle Winder.

Quill Winders

Ply yarns for fabric filling require a winding operation from bobbins or cones. This operation may be performed on a quiller. Single yarns on cones or tubes may also be wound on the quiller, rather than being spun directly on filling bobbins. Often, labor savings can be effected through increased battery hand assignments by using wound quills which result in increased yardage and reduced handling time per quill.

The calculation of spindle assignments for the quiller is considerably more complicated than for other equipment due to the number of variables that may be present. About the only element that can be considered as uniform when determining spindle stopped time is the quill transfer. Stopped time waiting at transfer breaks can only be accurately determined from time studies. After assignments have been set up on the basis of present elements of the job, the expected savings from the use of automatic bobbin loaders, magazine creels, and other refinements, may be evaluated.

In order to calculate the spindle efficiency take the total minutes per 8 hours and deduct the time allowances for personal purposes and machine cleaning. Next, calculate the stopped time for quill transfer, recreeling, repairing end breaks, and transfer breaks, by using the minutes available after personal and cleaning allowances as the basis for the calculations. Deduct the total stopped time from 480 minutes per day and divide the result by 480 minutes. Round off the answer for standard efficiency.

Determine the tenders' working minutes per 1000 quills and divide the time into the minutes available for winding. The result is the number of quills handled per day. Quills per tender per day divided by quills per 8 hour spindle yields the number of spindles per tender.

Payment for quillers may be by an hourly rate, or at a rate per 100 or 1000 quills, or per 100 pounds of yarn creeled. The latter method requires some estimating by the supervisor at shift changes to determine the quantity of yarn remaining in the creels.

Knot-Tieing Machine Operator

Although it is comparatively easy to make time studies and calculate job loads for the operator of the portable knot-tieing machine, it is practically impossible to predetermine his actual daily work requirements. The reason for this is that while a standard assignment may be based, for example, on tieing-in 17 loom beams in an 8 hour shift, some days only 10 or 12 beams run out and the next day as many as 24 may run out. On this basis the operator does not have a full assignment one day and is unable to complete his work on the second.

In order to overcome this condition, some mills would combine the job with that of loom overhauler or utility man, crediting time to knot-tieing only when this work is being done, and providing a spare machine which can be operated when needed. Under these conditions it is often not practical to pay by piece rates, although they can be used for standard cost purposes.

Weavers

In order to determine weavers' loom assignments, a considerable number of time studies are needed, especially if a variety of styles are run. The studies should include loom stops and the cause. By recording the pick clock production data before and after the study periods, and checking the loom speeds with a tachometer, loom efficiencies during the study may be determined, and the loom stops adjusted to the number of stops per active loom hour. Later when standard loom efficiencies are determined, the loom stops may be adjusted to stops per loom hour at standard efficiency.

Because a weaver's job is to keep the looms running and producing as much good cloth as possible, it is necessary to allow some time to patroling and inspecting. Many potential loom stops are eliminated by these inspection tours which allow the weaver to straighten out tangled ends on loom beams and remove stray ends from the warps and quill batteries. This patroling may be expressed either as minutes per 8 hour loom or as a percentage of the weaver's time.

It is necessary that the time study engineer be kept advised when changes or refinements are made affecting the running condition of the looms, so that new time studies can be made and adjustments in assignments effected accordingly.

Battery Hands

Looms assignments for battery hands depend primarily on the yards per quill. The more yards per quill for a given loom speed and fabric style, the fewer quills are required per loom hour and more looms can be assigned.

Time studies must be made to determine the battery hands' proper work pace, in terms of quills per hour. Tables may be set up for various quill sizes for both spun filling and wound quills, with extra walking allowances for longer-than-normal spaces between looms, securing supply of quills, and stripping waste from the batteries.

When full assignments are not possible, mills usually pay battery hands at an hourly rate, assigning other work such as quill-stripping or yarn hauling to round out a job. When a large number of styles are in each set, it may be better for the payroll department if this job is paid on the basis of a rate per battery-hour. This rate can be calculated by dividing the expected hourly, or 8 hour, earnings by the number of suggested looms per assignment.

CHAPTER 3

Maintenance of Time Standards

It is not necessary to discuss the need for standard maintenance since this is a *must* for the proper mill application of incentives. The following discussion pertains to the methods used in a typical plant to accomplish the sound and economical maintenance of time standards.

Standards Based on Mill Processing Specifications

The production machinery is run upon orders of the scheduling department, using processing specifications authorized by the technical department. These specifications indicate the product supplied to the production machines, the machine setups for proper operation, and the finished product desired from the process. Thus, a spinning mill processing specification shows the roving to be used for a particular yarn number, the speeds, gears, traveler, ring size, and other identifications required on the spinning frame, and the finished yarn number, twist multiple, and cotton dye desired.

Time standards are established from predetermined stopwatch time study values for practically all of the production machine operations. The workup of these standards is purposely arranged so that the information given on the processing specifications indicates which time study values are to be used to calculate the time standard. Thus, whenever a specification is issued or revised in any way, the time study department must originate or check the standard rates applicable to make sure correct wage payment is made. Payroll clerks receive their information with regard to code or specification numbers, and when they use the standards that agree with these code numbers they are up-to-date as regards all the changes involved.

Through daily checks with the specifications and specification changes issued, it is easy to correct standards. Since this job has to be done eventually, it is best to keep abreast of it daily. This prevents an overload of work on the time study department, and provides correct wage payment for the jobs so controlled. Examples of jobs covered in this category are frame tenders, roving markers, spinners, spooler and warper tenders, twister tenders, creelers and doffers, weavers and battery hands, and cloth inspectors.

Standards Not Based on Mill Processing Specifications

This group of standards is harder to maintain since it covers such variable and fluctuating items as end breaks and loom stops. As in the other classifications, these standards are also set up on stopwatch time study values. However, since these operations cannot be covered by mill processing specifications, it is more difficult to recognize when a change has occurred which will affect wage payment. Examples of jobs covered in this category are winder and respooler operators, packers and shippers, and sweepers and cleaners.

Methods changes, short cuts, gadgets, and end weather brekage, all may contribute to the loosening of establishedt staandards. Many of these changes would not normally be brough to the attention of the time study department. On the other hand, those changes which tend to tighten a standard are speedily brought to the attention of the time study men. In an attempt to keep up with all the factors that tend to influence work and hence wage payment, different methods are employed. Often, a work simplification program is carried on side by side with a suggestion system. One program is for salaried supervisors, the other for wage operators. Employees are encouraged to turn in their ideas for methods improvement, waste and cost reduction, and safety. When the plant committees meet to review the ideas turned in, all areas affected should be considered. If time study values need revision, the changes should be made when the proposal, suggestion, or idea is adopted and put into use.

End breakage, loom stops, cloth defects, and package weights need rechecking periodically. It is not always convenient to take the time to recheck all of these items on a regularly scheduled interval. However, if the mill is constantly adding to its equipment, changing the equipment it has, or is making more diversified products, it often happens that some standards become inapplicable and rechecking becomes necessary. At these times it is desirable to obtain not only the new information desired, but also to recheck existing information on portions of the job where no changes have been made. For example, one should check spinning end breakage on a new cotton mix, although part of the spinner's job is on a regular mix. Any noticeable change from standard end breakage would be detected at this time. As a further example, if a new type of loom is installed, with the weaver continuing to run some existing looms, data should be collected simultaneously on loom stops for both machines.

In any busy organization there is usually more work than can be done in a given period of time. This also holds true in a time study department. It must then be determined which important job comes first. To different people, different parts of the job are of the utmost importance. To help sift the jobs by importance, one may rely on various prior data and reports indicative of relative priorities of problems.

Reports

Monthly budgetary control reports show for each of the major manufacturing areas how much money was actually spent in the prior month, broken down by various job categories. A budget or standard is determined, and the variance over or under the standard is shown for each job category. Expenditures are also compared with the year-to-date actual budget. If a monthly figure appears to be unusual in any direction it is questioned, and occasionally a questioned figure leads to a change in job methods. Attention is given to correcting the situation.

A weekly labor analysis report shows the labor losses in dollars by departments. Labor losses, such as learners' guarantees, incorrect base rates, and failure to earn minimum wages are listed, and if these vary excessively they must be accounted for. The number and percentage of direct labor hours not on standard is shown. These data should reflect the activity of the time study department in keeping up with new jobs and new specifications and coverage of old jobs going on standard.

The average hourly earnings report shows, by major departments and shifts, the average hours worked and average straight time and overtime rate. Changes in average hourly rates must be scrutinized carefully to find the cause. If standards are at fault they must be changed.

The excess tally report on specially selected jobs, not paid on a hank, pick, or pound basis, is circulated to reflect the weekly loss incurred by incorrect counts or weighings. Better accuracy in reporting is always sought.

Many other reports indicate a trend of accomplishment. Each report serves a particular usefulness. By analyzing these reports, both individually and collectively, one can find weak spots in need of strengthening, as well as firm spots upon which to add to the soundness of the system.

CHAPTER 4

Frequency of Elemental Occurrences

Introduction

The dictionary definition of *frequency* is "the ratio of the number of occurrences of an event to the number of possible occurrences (in the same time)." It is apparent that however correct the definition may be in general, it is not sufficiently specific to be of much assistance in a discussion of frequencies in the textile industry. A good plan in such a discussion is to start with the concrete and familiar before introducing general principles, which is the procedure adopted herein.

Simple Example

Consider, as an example, the simple operation of yarn winding. Production proceeds at the rate of 1 ounce per minute, filling a cone of 60 ounces of yarn, when complete, from bobbins containing 5 ounces of yarn each. From the data it is a simple matter to derive the following:

1. The time required to wind a cone is 60 minutes of spindle running time, or as generally stated, 1 spindle running hour, SRH.

2. When the cone is full it is removed, and an empty tube is placed on the spindle for the next cone. This, therefore, occurs once per SRH, or in other words, the frequency of cone renewal per SRH is 1.

3. When a bobbin is empty, the cardboard tube is removed and a full bobbin is put in its place. Since 12, 5 ounce bobbins make one 60 ounce cone, this occurs 12 times per SRH, and thus, the frequency of bobbin renewal per SRH is 12.

To illustrate how frequencies are used we may assume that the operations are measured by stopwatch, and the following elemental times result: cone renewal, 0.24 minutes per occurrence, bobbin renewal, 0.15 minutes per occurrence. Then, it would follow that the operator time per SRH consisting of cone and bobbin renewals, is as shown in Table 4-1.

TABLE 4-1

Element	Frequency per SRH	Minutes per occurrence	Minutes per SRH[a]
Cone renewal	1	0.24	0.24
Bobbin renewal	12	0.15	1.80
Total operator time per SRH			2.04

[a] Frequency × minutes per occurrence.

These data may be used in workload and standard time determinations.

Considerations in Establishing Frequencies

The example shown illustrates three significant points concerning frequencies.

1. In some cases frequencies may be found by simple arithmetic.

2. The frequency, as well as the elemental time of each operation performed, must be known before a time standard can be completed.

3. All frequencies in any given case must be based on one and the same unit of production, or time interval. In the example above the SRH was chosen.

The reference or bench mark is a matter of choice or preference. Instead of using the SRH it would be equally valid to use the bobbin. This would be done if the standard time were to be established as so many minutes per bobbin wound. Then, the frequency of bobbin change per bobbin wound would be 1, and the frequency of cone change per bobbin wound would be $^1/_{12}$. Thus, frequencies do not have to be whole numbers. The operator time per bobbin wound would then be established as given in Table 4-2.

TABLE 4-2

Element	Frequency per bobbin	Minutes per occurrence	Minutes per bobbin[a]
Cone renewal	$^1/_{12}$	0.24	0.02
Bobbin renewal	1	0.15	0.15
Total operator time per bobbin			0.17

[a] Frequency × minutes per occurrence.

Again, these results would be used for determining the number of spindles to be assigned to one operator, and in calculating the standard time per bobbin. The frequencies determined in this way, whether per SRH or per bobbin, are as accurate as the basic data provided. Variations in bobbin weight, cone weight, or winding speed would introduce errors in computation. These variations should always be evaluated as far as possible and practicable, and be stated in conjunction with the derived results.

Expanded Example

The example given above omits mention of yarn breakages and is thus *idealized* for illustrative purposes. To make it more realistic we may assume that a study has been made of the rate of occurrence of end breaks, with the result that they appear to occur on the average 7 times per 12 bobbins wound in 1 SRH. The average frequency of yarn breakage per SRH would then be 7. This frequency, expressed in terms of bobbins wound, is equal to $^7/_{12}$. Let us now assume that 0.36 minutes represents the average time required to repair a break and restart the winding spindle. By combining this time with the frequency, the results shown in Table 4-3 are obtained.

TABLE 4-3

Element	Minutes per occurrence	Frequency per SRH	Minutes per SRH	Frequency per bobbin	Minutes per bobbin
Cone renewal	0.24	1	0.24	$^1/_{12}$	0.02
Bobbin renewal	0.15	12	1.80	1	0.15
Break repair	0.36	7	2.52	$^7/_{12}$	0.21
Total operator time			4.56		0.38

The operator time per bobbin wound would be 0.38 minutes, and 12 times this value, or 4.56 minutes, per SRH. One is equivalent to the other, and the form that is used depends on the general plan of the investigation.

The yarn breakage above supplies an example of a frequency that cannot be calculated. A break is a random and unpredictable event, and thus, there is no basis for such a calculation. To evaluate the frequency it is necessary to study the course of events over a period, and, after making certain assumptions, project them into the unknown future.

There is little more that can be said on the subject of frequency that does not involve consideration of the error contained in the various estimates which will now be considered. *Error* refers to the range of values within which the truth is expected to be found, with a certain degree of probability.

Error and Probability

The time to change a cone on a winding spindle was measured with a stopwatch on 10 occasions and the times recorded were: 0.18, 0.30, 0.27, 0.21, 0.25, 0.23, 0.20, 0.19, 0.30, and 0.27 minutes.

The average time, obtained in the usual way, is 0.24, and this is the figure previously given (Table 4-1) as the time for the cone changing operation. It is now seen, however, that the actual measured time varied from 0.18 to 0.30 minutes, and a reading of 0.24 minutes was never actually observed. The statistician introduces the idea of a population of readings for the operation time, and would refer to the above as a sample of 10 readings. (Incidentally, he would usually demand a larger sample consisting of at least 30 or 40 readings, before drawing any conclusions. To keep the arithmetic simple, however, the calculations are continued on the basis of 10 readings.) He would say the sample average is likely to be a guide to the population average, which is the figure sought. In fact, since a lifetime would be too short to determine the true population average, he will say the sample average can be used as an approximation, but it is necessary to name the probable magnitude of error involved. The fairly simple principles involved, which will be found in almost any text on statistics, lead to the following:

$$N_o = (100t\sigma/E\overline{X})^2$$

where N_o indicates the number of observations or readings of the time of the element under study, $\overline{X}$ indicates the average time derived from the sample of readings of the element, and σ denotes the standard deviation.* E represents the percentage error in the evaluation of $\overline{X}$. When E is, for example, equal to 5, this means that the true time value of the element lies within ± 5 per cent of the average value, $\overline{X}$, derived from the sample. The value t is a factor which takes into account the probability. Table 4-4 contains a listing of t factors and their corresponding probabilities (also the betting interpretation).

TABLE 4-4

t	Probability, %	Betting odds	t	Probability, %	Betting odds
1.00	68.3	2.2	1.80	92.8	12.9
1.20	77.0	3.3	2.00	95.5	21.0
1.45	83.8	5.2	2.50	98.8	79.5
1.50	86.6	6.5	3.00	99.7	383.0

The equation for N_o supplies the answer to the question: How many observations are needed to establish the value of $\overline{X}$ within $\pm E$ per cent with a probability corresponding to t?

* N. L. Enrick, *Time Study Manual for the Textile Industry*, Textile Book Publishers (Interscience), New York, 1960.

TABLE 4-5
Number of Tests Required for a 0.3% Sampling Risk[a]

	Allowable sampling error, %E									
% V	0.5	0.8	1.0	1.2	1.5	2.0	2.5	3.0	3.5	4.0
0.5	9	4	2	2	1	1	1	1	1	1
0.6	13	5	3	2	2	1	1	1	1	1
0.7	18	7	4	3	2	1	1	1	1	1
0.8	23	9	6	4	3	1	1	1	1	1
1.0	36	14	9	6	4	2	1	1	1	1
1.2	52	20	13	9	6	3	2	1	1	1
1.4	71	28	18	12	8	4	3	2	1	1
1.6	92	36	23	16	10	6	4	3	2	2
1.8	117	46	29	20	13	7	5	3	2	1
2.0	144	56	36	25	16	9	6	4	3	2
2.2	174	68	44	30	19	11	7	5	4	3
2.4	207	81	52	36	23	13	8	6	4	3
2.6	243	95	61	42	27	15	10	8	5	4
2.8	282	110	71	49	31	18	11	8	6	4
3.0	324	127	81	56	36	20	13	9	7	5
3.2	369	144	92	64	41	23	15	10	8	6
3.4	416	163	104	72	46	26	17	12	8	7
3.6	467	182	117	81	52	29	19	13	10	7
3.8	520	203	130	90	58	32	21	14	11	8
4.0	576	225	144	100	64	36	23	16	12	9
4.2	635	248	159	110	71	40	25	18	13	10
4.4	697	272	174	121	77	44	28	19	14	11
4.6	762	298	190	132	85	48	30	21	16	12
4.8	829	324	207	144	92	52	33	23	17	13
5.0	900	352	225	156	100	56	36	25	18	14
5.2	973	380	243	169	108	61	39	27	20	15
5.4	1,050	410	262	182	117	66	42	29	21	16
5.6	1,129	441	282	196	125	71	45	31	23	18
5.8	1,211	473	303	210	135	76	48	34	25	19
6.0	1,296	506	324	225	144	81	52	36	26	20
7.0	1,764	689	441	306	196	110	71	49	36	28
8.0	2,304	900	576	400	256	144	92	64	47	36
10.0	3,600	1,406	900	625	400	225	144	100	73	56
12.0	5,184	2,025	1,296	900	576	324	207	144	106	81
15.0	8,100	3,164	2,025	1,406	900	506	324	225	165	127
20.0	14,400	5,625	3,600	2,500	1,600	900	576	400	294	225
25.0	22,500	8,789	5,625	3,906	2,500	1,406	900	625	459	351
30.0	32,400	12,656	8,100	5,625	3,600	2,025	1,297	900	661	506
35.0	44,100	17,227	11,025	7,656	4,900	2,756	1,764	1,225	900	689
40.0	57,600	22,500	14,400	10,000	6,400	3,600	2,304	1,600	1,176	900

Source: N. L. Enrick, *Statistics Course Notes*, 1960.

[a] Example: *Given*—$V = 3\%$, $E = 4\%$. *Find*—the number of tests required. Enter the table at the 3% level and proceed horizontally until the column for $\%E = 4$ is reached—thus, 5 tests are required.

Other things being equal, N will vary in proportion to t^2. If, for instance, N_o is 100 when t is 1 (68 per cent probability), N_o will equal 400 when t is 2 (95.5 per cent probability) and 900 when t is 3 (99.7 per cent probability). Whether the additional confidence is worth the additional work is a matter to be decided by

TABLE 4-5 (*continued*)

Allowable sampling error, %E													
4.5	5.0	5.5	6.0	7.0	8.0	10.0	12.0	15.0	18.0	20.0	25.0	30.0	%V
1	1	1	1	1	1	1	1	1	1	1	1	1	0.5
1	1	1	1	1	1	1	1	1	1	1	1	1	0.6
1	1	1	1	1	1	1	1	1	1	1	1	1	0.7
1	1	1	1	1	1	1	1	1	1	1	1	1	0.8
1	1	1	1	1	1	1	1	1	1	1	1	1	1.0
1	1	1	1	1	1	1	1	1	1	1	1	1	1.2
1	1	1	1	1	1	1	1	1	1	1	1	1	1.4
1	1	1	1	1	1	1	1	1	1	1	1	1	1.6
1	1	1	1	1	1	1	1	1	1	1	1	1	1.8
2	1	1	1	1	1	1	1	1	1	1	1	1	2.0
2	2	1	1	1	1	1	1	1	1	1	1	1	2.2
3	2	2	1	1	1	1	1	1	1	1	1	1	2.4
3	2	2	2	1	1	1	1	1	1	1	1	1	2.6
3	3	2	2	1	1	1	1	1	1	1	1	1	2.8
4	3	3	2	2	1	1	1	1	1	1	1	1	3.0
5	4	3	3	2	1	1	1	1	1	1	1	1	3.2
5	4	3	3	2	2	1	1	1	1	1	1	1	3.4
6	5	4	3	2	2	1	1	1	1	1	1	1	3.6
6	5	4	4	3	2	1	1	1	1	1	1	1	3.8
7	6	5	4	3	2	1	1	1	1	1	1	1	4.0
8	6	5	4	3	2	2	1	1	1	1	1	1	4.2
9	7	6	5	4	3	2	1	1	1	1	1	1	4.4
9	8	6	5	4	3	2	1	1	1	1	1	1	4.6
10	8	7	6	4	3	2	1	1	1	1	1	1	4.8
11	9	7	6	5	4	2	2	1	1	1	1	1	5.0
12	10	8	7	5	4	2	2	1	1	1	1	1	5.2
13	10	9	7	5	4	3	2	1	1	1	1	1	5.4
14	11	9	8	6	4	3	2	1	1	1	1	1	5.6
15	12	10	8	6	5	3	2	1	1	1	1	1	5.8
16	13	11	9	7	[illegible]	3	2	1	1	1	1	1	6.0
22	18	15	12	9	7	4	3	2	1	1	1	1	7.0
28	23	19	16	12	9	6	4	3	2	1	1	1	8.0
44	36	30	25	18	14	9	6	4	3	2	1	1	10.0
64	52	43	36	26	20	13	9	6	4	3	2	1	12.0
100	81	67	56	41	32	20	14	9	6	5	3	2	15.0
178	144	119	100	73	56	36	25	16	11	9	6	4	20.0
278	225	186	156	115	88	56	39	25	17	14	9	6	25.0
400	324	268	225	165	127	81	56	36	25	20	13	9	30.0
544	441	364	306	225	172	110	77	49	34	28	18	12	35.0
711	576	476	400	294	225	144	100	64	44	36	23	16	40.0

the merits of each case. In time studies the most generally adopted probability is the one corresponding to t is 2, which is approximately 95 per cent.

Again, other things being equal, N_o will vary inversely as the square of E. If it requires 400 readings to hold the probable error to ± 5 per cent, it will require only 100 if the wider range of ± 10 per cent can be tolerated for the same degree of confidence or probability level.

TABLE 4-6
Number of Tests Required for a 1.0% Sampling Risk[a]

	Allowable sampling error, %*E*									
%*V*	0.5	0.8	1.0	1.2	1.5	2.0	2.5	3.0	3.5	4.0
0.5	7	3	2	1	1	1	1	1	1	1
0.6	10	4	2	2	1	1	1	1	1	1
0.7	13	5	3	2	1	1	1	1	1	1
0.8	17	7	4	3	2	1	1	1	1	1
1.0	27	11	7	5	3	2	1	1	1	1
1.2	39	15	10	7	4	2	2	1	1	1
1.4	53	21	13	9	6	3	2	1	1	1
1.6	65	27	17	12	7	4	3	2	1	1
1.8	88	34	22	15	10	5	4	2	2	1
2.0	108	42	27	19	12	7	4	3	2	2
2.2	131	51	33	23	15	8	5	4	3	2
2.4	156	61	39	27	17	10	6	4	3	2
2.6	183	71	46	32	20	11	7	5	4	3
2.8	212	83	53	37	24	13	8	6	4	3
3.0	243	95	61	42	27	15	10	7	5	4
3.2	277	108	69	48	31	17	11	8	6	4
3.4	313	122	78	54	35	20	13	9	6	5
3.6	350	137	88	61	39	22	14	10	7	5
3.8	390	153	98	68	43	24	16	11	8	6
4.0	433	169	108	75	48	27	17	12	9	7
4.2	477	186	119	83	53	30	19	13	10	7
4.4	523	204	131	91	58	33	21	15	11	8
4.6	572	224	143	99	64	36	23	16	12	9
4.8	623	243	156	108	69	39	25	17	13	10
5.0	676	264	169	117	75	42	27	19	14	11
5.2	731	286	183	127	81	46	29	20	15	11
5.4	788	308	197	137	88	49	32	22	16	12
5.6	848	331	212	147	94	53	34	24	17	13
5.8	910	355	227	158	101	57	36	25	19	14
6.0	973	380	243	169	108	61	39	27	20	15
7.0	1,325	518	331	230	147	83	53	37	27	31
8.0	1,731	676	433	300	192	108	69	48	35	27
10.0	2,704	1,056	676	469	300	169	108	75	55	42
12.0	3,894	1,521	973	676	433	243	156	108	79	61
15.0	6,084	2,377	1,521	1,056	675	380	243	169	124	95
20.0	10,816	4,225	2,704	1,878	1,201	676	433	300	220	169
25.0	16,900	6,602	4,225	2,934	1,876	1,056	676	469	345	264
30.0	24,336	9,506	6,084	4,225	2,701	1,521	973	675	496	380
35.0	33,124	12,939	8,281	5,750	3,677	2,070	1,325	919	676	518
40.0	43,264	16,900	10,816	7,511	4,802	2,704	1,731	1,201	882	676

Source: N. L. Enrick, *Statistics Course Notes*, 1960.
[a] Example: *Given*—$V = 4\%$, $E = 5\%$. *Find*—the number of tests required. Enter the table at 4% level and proceed horizontally until the column for $\%E = 5$ is reached—thus, 4 tests are required.

Sometimes the question is in a different form. After taking *n* observations, what is the probable range of error? Put *n* in place of N_o in the above formula and transpose the terms, to yield *E* as the required answer:

$$E = (100t\sigma)/(\bar{X}\sqrt{n})$$

TABLE 4-6 (*continued*)

Allowable sampling error, %E													
4.5	5.0	5.5	6.0	7.0	8.0	10.0	12.0	15.0	18.0	20.0	25.0	30.0	%V
1	1	1	1	1	1	1	1	1	1	1	1	1	0.5
1	1	1	1	1	1	1	1	1	1	1	1	1	0.6
1	1	1	1	1	1	1	1	1	1	1	1	1	0.7
1	1	1	1	1	1	1	1	1	1	1	1	1	0.8
1	1	1	1	1	1	1	1	1	1	1	1	1	1.0
1	1	1	1	1	1	1	1	1	1	1	1	1	1.2
1	1	1	1	1	1	1	1	1	1	1	1	1	1.4
1	1	1	1	1	1	1	1	1	1	1	1	1	1.6
1	1	1	1	1	1	1	1	1	1	1	1	1	1.8
1	1	1	1	1	1	1	1	1	1	1	1	1	2.0
2	1	1	1	1	1	1	1	1	1	1	1	1	2.2
2	2	1	1	1	1	1	1	1	1	1	1	1	2.4
2	2	2	1	1	1	1	1	1	1	1	1	1	2.6
3	2	2	1	1	1	1	1	1	1	1	1	1	2.8
3	2	2	2	1	1	1	1	1	1	1	1	1	3.0
3	3	2	2	1	1	1	1	1	1	1	1	1	3.2
4	3	3	2	2	1	1	1	1	1	1	1	1	3.4
4	4	3	2	2	1	1	1	1	1	1	1	1	3.6
5	4	3	3	2	2	1	1	1	1	1	1	1	3.8
5	4	4	3	2	2	1	1	1	1	1	1	1	4.0
6	5	4	3	2	2	1	1	1	1	1	1	1	4.2
6	5	4	4	3	2	1	1	1	1	1	1	1	4.4
7	6	5	4	3	2	1	1	1	1	1	1	1	4.6
8	6	5	4	3	2	2	1	1	1	1	1	1	4.8
8	7	6	5	3	3	2	1	1	1	1	1	1	5.0
9	7	6	5	4	3	2	1	1	1	1	1	1	5.2
10	8	7	5	4	3	2	1	1	1	1	1	1	5.4
10	8	7	6	4	3	2	1	1	1	1	1	1	5.6
11	9	8	6	5	4	2	2	1	1	1	1	1	5.8
12	10	8	7	5	4	2	2	1	1	1	1	1	6.0
16	13	11	9	7	5	3	2	1	1	1	1	1	7.0
21	17	14	12	9	7	4	3	2	1	1	1	1	8.0
33	27	22	19	14	11	7	5	3	2	2	1	1	10.0
48	39	32	27	20	15	10	7	4	3	2	2	1	12.0
75	61	50	42	31	24	15	10	7	5	4	2	2	15.0
134	108	89	75	55	42	27	19	12	8	7	4	3	20.0
209	169	139	117	86	66	42	29	19	13	11	7	5	25.0
301	243	201	169	124	95	61	42	27	19	15	10	7	30.0
409	331	273	230	169	129	82	57	37	26	21	13	9	35.0
534	433	357	301	221	169	108	75	48	34	27	17	12	40.0

Note that n is the number of readings already taken, whereas N_o is the number that should be taken, including the n already obtained.

The data in Tables 4-5 to 4-7 provide a ready reference for the determination of the proper sample sizes required for various practical conditions. The tables are for risks of error of 0.3, 1.0 and 5.0 per cent. These risks correspond to probability levels of assurance of 99.7, 99.0 and 95.0 per cent. A wide range of values

TABLE 4-7
Number of Tests Required for a 5% Sampling Risk[a]

%V	Allowable sampling error, %E									
	0.5	0.8	1.0	1.2	1.5	2.0	2.5	3.0	3.5	4.0
0.5	4	2	1	1	1	1	1	1	1	1
0.6	6	2	1	1	1	1	1	1	1	1
0.7	8	3	2	1	1	1	1	1	1	1
0.8	10	4	3	2	1	1	1	1	1	1
1.0	16	6	4	3	2	1	1	1	1	1
1.2	23	9	6	4	3	1	1	1	1	1
1.4	31	12	8	5	3	2	1	1	1	1
1.6	41	16	10	7	5	3	2	1	1	1
1.8	52	20	13	9	6	3	2	1	1	1
2.0	64	25	16	11	7	4	3	2	1	1
2.2	77	30	19	13	9	5	3	2	2	1
2.4	92	36	23	16	10	6	4	3	2	1
2.6	108	42	27	19	12	7	4	3	2	2
2.8	125	49	31	22	14	8	5	3	3	2
3.0	144	56	36	25	16	9	6	4	3	2
3.2	164	64	41	28	18	10	7	5	3	3
3.4	185	72	46	32	21	12	7	5	4	3
3.6	207	81	52	36	23	13	8	6	4	3
3.8	231	90	58	40	26	14	9	6	5	4
4.0	256	100	64	44	28	16	10	7	5	4
4.2	282	110	71	49	31	18	11	8	6	4
4.4	310	121	77	54	34	19	12	9	6	5
4.6	339	132	85	59	38	21	14	9	7	5
4.8	369	144	92	64	41	23	15	10	8	6
5.0	400	156	100	69	44	25	16	11	8	6
5.2	433	169	108	75	48	27	17	12	9	7
5.4	467	182	117	81	52	29	19	13	10	7
5.6	502	196	125	87	56	31	20	14	10	8
5.8	538	210	135	93	60	34	22	15	11	8
6.0	576	225	144	100	64	36	23	16	12	9
7.0	784	306	196	136	87	49	31	22	16	12
8.0	1,024	399	256	178	114	64	41	28	21	16
10.0	1,600	624	400	278	178	100	64	44	33	25
12.0	2,304	899	576	400	256	144	92	64	47	36
15.0	3,600	1,404	900	625	400	225	144	99	73	56
20.0	6,400	2,496	1,600	1,110	710	400	256	178	131	100
25.0	10,000	3,900	2,500	1,735	1,110	625	400	278	204	156
30.0	14,400	5,616	3,600	2,498	1,598	900	576	400	294	225
35.0	19,600	7,644	4,900	3,401	2,176	1,225	784	544	400	306
40.0	25,600	9,984	6,400	4,442	2,842	1,600	1,024	711	522	400

Source: N. L. Enrick, *Statistics Course Notes*, 1960.
[a] Example: *Given*—$V = 4\%$, $E = 2\%$. *Find*—the number of tests required. Enter the table at the 4% level and proceed horizontally until the column for $\%E = 2$ is reached—thus, 16 tests are required.

of per cent E is shown along the top of each table. The measure of variability used is the variation coefficient, V, which is the value of the standard deviation expressed as a percentage of the average, $\bar{X}$. Although there is an infinitely wide range of standard deviations that may be encountered, the ratio of such standard

TABLE 4-7 (*continued*)

Allowable sampling error, %E													
4.5	5.0	5.5	6.0	7.0	8.0	10.0	12.0	15.0	18.0	20.0	25.0	30.0	%V
1	1	1	1	1	1	1	1	1	1	1	1	1	0.5
1	1	1	1	1	1	1	1	1	1	1	1	1	0.6
1	1	1	1	1	1	1	1	1	1	1	1	1	0.7
1	1	1	1	1	1	1	1	1	1	1	1	1	0.8
1	1	1	1	1	1	1	1	1	1	1	1	1	1.0
1	1	1	1	1	1	1	1	1	1	1	1	1	1.2
1	1	1	1	1	1	1	1	1	1	1	1	1	1.4
1	1	1	1	1	1	1	1	1	1	1	1	1	1.6
1	1	1	1	1	1	1	1	1	1	1	1	1	1.8
1	1	1	1	1	1	1	1	1	1	1	1	1	2.0
1	1	1	1	1	1	1	1	1	1	1	1	1	2.2
1	1	1	1	1	1	1	1	1	1	1	1	1	2.4
1	1	1	1	1	1	1	1	1	1	1	1	1	2.6
2	1	1	1	1	1	1	1	1	1	1	1	1	2.8
2	1	1	1	1	1	1	1	1	1	1	1	1	3.0
2	2	1	1	1	1	1	1	1	1	1	1	1	3.2
2	2	2	1	1	1	1	1	1	1	1	1	1	3.4
3	2	2	1	1	1	1	1	1	1	1	1	1	3.6
3	2	2	2	1	1	1	1	1	1	1	1	1	3.8
3	3	2	2	1	1	1	1	1	1	1	1	1	4.0
3	3	2	2	1	1	1	1	1	1	1	1	1	4.2
4	3	3	2	2	1	1	1	1	1	1	1	1	4.4
4	3	3	2	2	1	1	1	1	1	1	1	1	4.6
5	4	3	3	2	1	1	1	1	1	1	1	1	4.8
5	4	3	3	2	2	1	1	1	1	1	1	1	5.0
5	4	4	3	2	2	1	1	1	1	1	1	1	5.2
6	5	4	3	2	2	1	1	1	1	1	1	1	5.4
6	5	4	3	3	2	1	1	1	1	1	1	1	5.6
7	5	4	4	3	2	1	1	1	1	1	1	1	5.8
7	6	5	4	3	2	1	1	1	1	1	1	1	6.0
10	8	6	5	4	3	2	1	1	1	1	1	1	7.0
13	10	8	7	5	4	2	1	1	1	1	1	1	8.0
20	16	13	11	8	6	4	3	2	1	1	1	1	10.0
28	23	19	16	12	9	6	4	3	2	1	1	1	12.0
44	36	30	25	18	14	9	6	4	3	2	1	1	15.0
79	64	53	44	33	25	16	11	7	5	4	3	2	20.0
124	100	83	70	51	39	25	17	11	8	6	4	3	25.0
178	144	119	100	73	56	36	25	16	11	9	6	4	30.0
242	196	162	136	100	76	49	34	22	15	12	8	5	35.0
316	256	211	178	131	100	64	44	28	20	16	10	7	40.0

deviations to the corresponding average is relatively limited in range. Thus, by using the variation coefficient, it was possible to develop these tables for a wide range of practical problems, probably including all situations encountered in frequency determination for time study work. The tables are, of course, also suitable for statistical quality control in determining the proper amount of control testing required.

CHAPTER 5

Frequency of Random Occurrences (Ends Down and Loom Stops)

Introduction

All operator tasks cannot be scheduled. Some must be performed as the need arises, because neither the time nor place of their occurrence can be predicted. Such random events include roving run outs, ends down, loom stops, split laps, chokes, and other faults and defects that interrupt production, or produce bad quality and require the operator's attention. Events of this nature are of random occurrence because they are associated with a number of cause factors, most of which are capricious in their effect. Any one factor or combination of factors may cause the undesired event to happen.

Such events are called *attributes* because they can be evaluated only by counting the frequency of occurrence. There is no recognized degree of variation in quality; either the event occurs or it does not. Because of these two characteristics, random occurrence and attribute classification, the pattern of their frequency tends to conform to either a binomial or Poisson-type distribution. In either case the statistical treatment is very similar.

Typical Distributions of Frequencies

Figure 5-1 shows a slightly modified binomial frequency distribution. It represents the expected frequency of spindles having 0, 1, 2, and more ends down during a 100 hour run, when the over-all rate is 40 end breaks per 1000 spindle hours. The percentage of spindles falling in each group is shown on the vertical

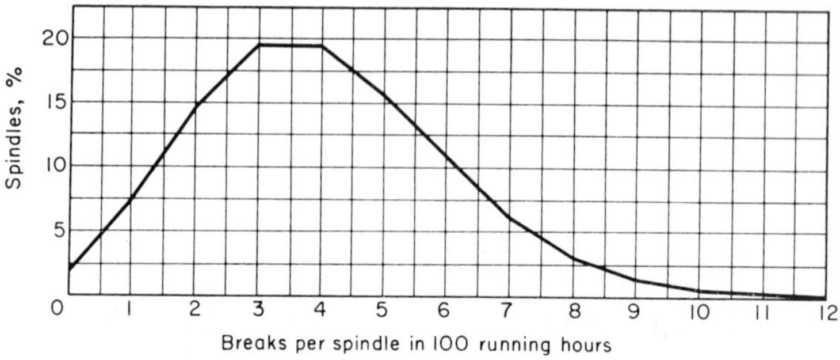

Figure 5-1. Frequency distribution of end breaks per spindle (expected) for a rate of 40 ends down per 1,000 spindle hours and 100 running hours of observation. ($p = 0.04$; $n = 100$; $pn = 4.0$; $\sigma = \sqrt{pn} = 2$; and $V = 50\%$.)

scale, whereas the number of breaks per spindle is given on the horizontal scale. From the graph and also from Table 5-1 it can be observed that about 2 per cent of the spindles are expected to have no breaks during a continuous 100 hour period, about 7 per cent are expected to have one break during a continuous 100 hour period, and approximately 15 per cent are expected to have

TABLE 5-1
Expected Distribution of End Breaks[a]

Number of breaks per spindle	Spindles having the number of breaks indicated, %
0	1.8
1	7.4
2	14.6
3	19.6
4	19.6
5	15.6
6	10.4
7	6.0
8	3.0
9	1.3
10	0.5
11	0.2
12	0.1

[a] Illustrated for the data in Figure 5-1, based on 40 ends down per 1000 spindle hours and an observation period of 100 running hours.

3 breaks. Actual inspection has confirmed this theoretical frequency rather closely.

Figure 5-2 shows the expected frequencies for shorter runs. When the inspection period is short, a rather high percentage of the spindles will have zero end breaks during the test. These are

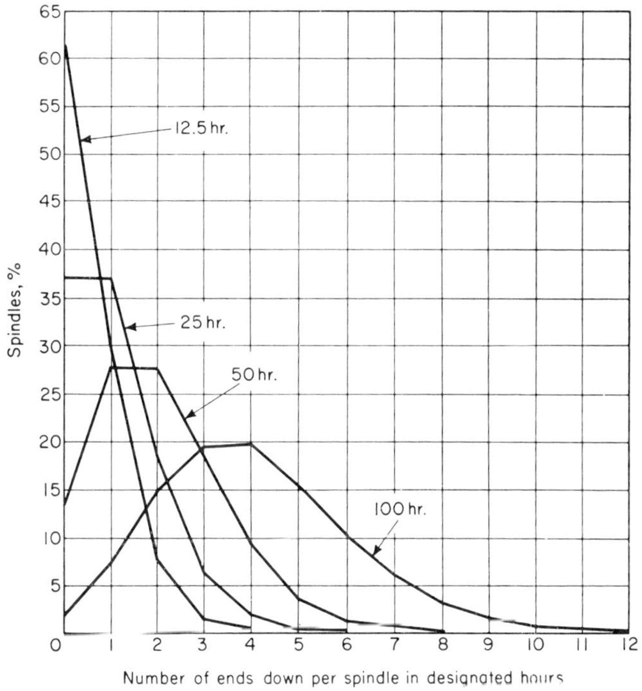

Figure 5-2. Frequency distribution of end breaks per spindle (expected) for a rate of 40 ends down per 1000 spindle hours for various periods of running time observations.

typical Poisson exponential series. Most attribute inspection test data, if represented by a frequency distribution, would closely resemble the curves.

In many instances we are concerned with the average number of events, attributes, defects, or work stops in a definite period of time, or specific sample size. When we are concerned with the average number of events or frequencies, the statistics shown in Table 5-2 are applicable.

Sampling and Reliability

Figure 5-3 shows how reliability of a sample estimate improves as the number of inspections is increased. This particular graph represents a spinner's job with a true average rate of 60 end breaks per hour. The per cent sampling error, $3 \times 100/\sqrt{N_2}$, is represented by the graduations on the vertical scale. Hours of inspection per test is shown on the horizontal scale. From the downward sweep of the graph it is very evident that per cent sample error decreases rapidly for the first few hours of inspection, and then soon reaches a point of diminishing returns. After several hours, the improvement in the estimate of true rate of end breakage hardly justifies the added hours of inspection required to attain it.

This graph is typical of all sampling inspection and testing. There is always a sample error, for any test short of 100 per cent inspection. We cannot ignore its existence. Our concern is that it be equal to or less than some practical differences that we are

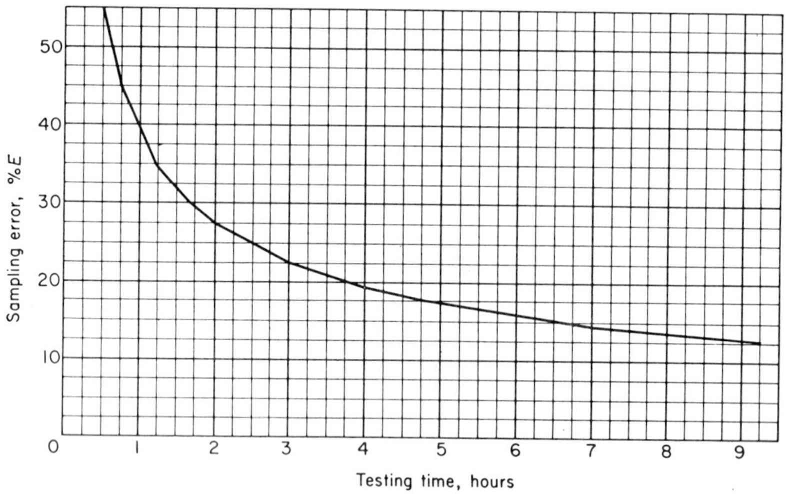

Figure 5-3. Relation of sampling error to hours of testing or observation time. The sampling error decreases with increasing inspection time, based on the following relationship: Error, $\% = 3 \times 100/\sqrt{N}$. The reliability of the inspection data follows the curve of diminishing return, as illustrated by the ends down on one spinner's job (average ends down per hour is 60).

TABLE 5-2
Statistical Quantities Useful in Evaluating Frequencies of Occurrence[a]

Classification	Average number	Standard deviation, σ	Variation coefficient, V
1. Binomial, $p > 0.1$	pn	$\sqrt{pn\,(1-p)}$	$100\sqrt{(1-p)/pn}$
2. Binomial, $p < 0.1$	pn	$\sqrt{pn}$	$100/\sqrt{pn}$
3. Poisson, $p < 0.1$	N	$\sqrt{N}$	$100/\sqrt{N}$

[a] p = probability—for example, for 40 ends down observed in 1000 spindle hours, $p = 40/1000 = 0.04$; n = total possible occurrences observed—in the example above, n = 1000 spindle hours; $N = pn$ = average occurrence—for the example above, $pn = 0.04 \times 1000 = 40$ ends down; and $V = (\sigma \times 100)/\text{average}$—thus, for lines 1, 2, and 3 above:

1. $V = (100\sqrt{pn(1-p)})/pn = 100\sqrt{(1-p)/pn}$
2. $V = 100\sqrt{pn}/pn = 100/\sqrt{pn}$
3. $V = (100\sqrt{N})/N = 100/\sqrt{N}$

seeking to measure. In the illustration just given, it is realized that in setting up a spinner's job some tolerance must be allowed for actual fluctuation in the rate of end breakages. Designating this tolerance as T, we can write this practical condition in the form of an equation:

$$T = t\sigma/\sqrt{n} \qquad \text{general equation} \tag{1}$$

$$\%T = t100/\sqrt{n} \qquad \text{specific equation} \tag{2}$$

where t is the factor associated with sample risk (explained in Chapter 4). Determination of the practical tolerance T can best be explained by Illustrative Example 5-1. For the example given, it is evident that the tolerance for a spinner's job of 12 sides is 60.5 ± 7.5 piece-ups per hour, or 40 ± 5 ends down per 1000 spindle hours. Per cent T is 100 × 5/40 equal to 12.5 per cent.

It is apparent that equation (2) can be rearranged to show:

$$N = (100t/\%T)^2 \tag{3}$$

where N represents the minimum number of events, occurrences, defects, or other attributes which must be observed in order that the sampling error, $\%E$, be equal to or smaller than the tolerance, $\%T$.

Sampling Risks

Figure 5-4 will help explain sample risk. When small samples are taken from three sources that differ slightly in quality, the overlapping of sample results constitutes a risk in making a

ILLUSTRATIVE EXAMPLE 5-1

Determination of Tolerance Limits as a Criterion for Changing Frame Side Assignments with Changes in Ends Down Rates in Spinning

Given

The spinner's job is based on 40 ends down per 1000 spindle hours. The standard job is 12 sides × 126 spindles = 1512 spindles. Piecing up ends accounts for 25.2% of the spinner's time.

Problem

When should the spinner's job be adjusted, because of a change in the rate of end breakage, by plus or minus one side?

Solution

1. For a standard of 12 sides, the average piece-ups are equal to 1512 × 40/1000 or 60.48 piece-ups per hour. Since 60.48 piece-ups require 25.2% of the spinner's time, one piece-up requires 0.41666%. Other duties on the 12 sides require 74.8% of the spinner's time, one side requiring 6.2333%. Thus,

12 (6.2333%) + 60.48 (0.41666%) = 100%

2. The equivalent work load with only 11 sides is as follows:

100% job load = 11(6.2333%) + ends down (0.41666%) × factor × piece-ups = 31.433% of a job, and the number of piece-ups or ends down is 31.433 ÷ 0.4166 = 75.4.

3. Equivalent work load with 13 sides is as follows:

Since 13 (6.2333%) + ends down (0.416665) = 100% job load

Therefore, 0.41666 × ends down = 18.9666% workload, and the number of piece-ups or ends down is 45.5.

Conclusion

11 Sides, standard ends down per hour	=	75.40
upper tolerance	=	67.94
12 Sides, standard ends down per hour	=	60.48
lower tolerance	=	52.99
13 Sides, standard ends down per hour	=	45.50

decision regarding their respective quality. If larger samples are taken the sample results would not overlap so frequently and less risk would be involved in making a decision.

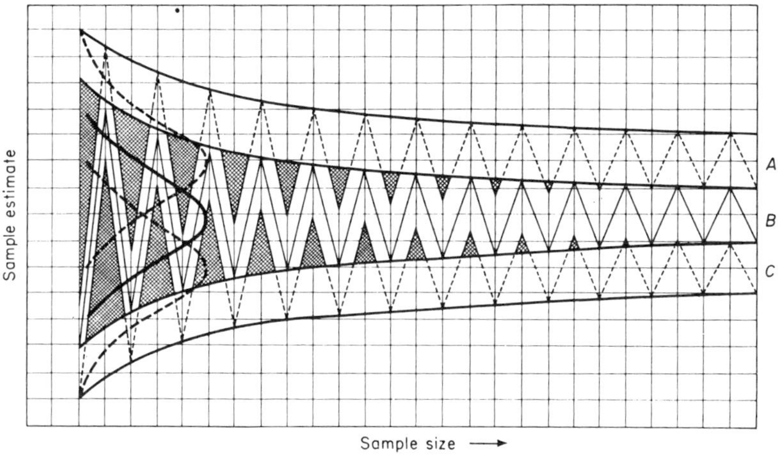

Figure 5-4. Since sample averages fluctuate about their true average, increasing sample sizes reduce these fluctuations. In this illustration, samples are taken from three slightly different quality levels *A*, *B*, and *C*. The sampling errors, indicated by the zigzag lines, decrease with increasing sample size. Shaded areas indicate the overlapping of tolerances for sampling error. The relative amount of shaded area represents the sampling risk.

Sample Calculations

The following examples will illustrate the use of equation (3) for determining the required amount of observations of defects. In the cases given here, a *defect* represents an end down in spinning.

For a 5 per cent sampling risk, t equals 1.96 and the $\% T$ is 12.5. The required ends down to be observed, N, would be:

$$N = (100 \times 1.96/12.5)^2 = 246 \qquad (4)$$

The figure 246 represents the minimum number of ends down to be observed. If, instead of a 5 per cent sampling risk (or a 95 per cent confidence level) we had used a risk of only 0.3 per cent, representing a 99.7 per cent confidence level, the value of t would have become 3. This value yields:

$$N = (100 \times 3/12.5)^2 = 576 \qquad (5)$$

It should be noted that, in place of the t of 1.96 used above, it is frequent practice to use a rounded out value of t equal to 2.

Application of Calculations

The application of the calculations just presented is illustrated further in Table 5-3. For a standard job consisting of 1,500 spindles, with a rate of 40 end breaks per 1,000 spindle hours, and various inspection times, the values are given for the lower and upper tolerance limits for observed ends down, based on a $\%T$ of ± 12.5, and the minimum and maximum limits for the sampling error, E, based on a sampling risk of 5 per cent. Thus,

TABLE 5-3
Relation of Tolerance Limits to Limits of Sampling Error, Showing Point Where Changes in Assignment Can Be Considered[a]

Inspection time, hours	Lower limit of job tolerance, T (-12.5% N)	Standard job at 40 ends down per 1000 spindle hours: $E_{min.} = N - 2\sqrt{N}$	Average, N	$E_{max.} = N + 2\sqrt{N}$	Upper limit of job tolerance, T ($+12.5\%$ N)
0.25	13	7	15	23	17
0.50	26	19	30	41	34
0.75	39	31	45	59	51
1.00	52	44	60	76	68
2.00	105	98	120	142	135
3.00	158	153	180	207	202
4.00	210	209	240	271	270
5.00	262	265	300	335	338
6.00	315	322	360	398	405
7.00	368	379	420	461	472
8.00	420	436	480	524	540

[a] At 5 hours inspection time, the limits of sampling error (columns 3 and 5) are narrower than the limits of the job tolerance (columns 2 and 6); thus, decisions affecting frame side assignments can be made after 5 hours, or more, of inspection.

for the average, N, of ends down observed during the inspection periods indicated, the minimum and maximum limits of E are given by $N \pm 2\sqrt{N}$. It will be noted from the table that for inspection times of less than 4 hours, the limits of E are wider than the limits of T; no statistical decision can be reached, therefore. At the 4 hour level, a borderline situation exists, and the two sets of limits coincide, approximately. After 5 hours the limits of E are definitely closer than those of T, permitting statisti-

TABLE 5-4
Minimum Number of Observations Required for Various Tolerances and Risks[a]

Job tolerance, %T	Sampling risk, % 0.1 (t = 3.29)	0.27 (t = 3.0)	1.0 (t = 2.575)	6.0 (t = 1.960)	10.0 (t = 1.645)
0.5	432,964	360,000	265,225	153,664	108,241
1.0	108,241	90,000	66,306	38,416	27,060
1.5	48,107	40,000	29,469	17,074	12,026
2.0	27,060	22,500	16,576	9,604	6,765
2.5	17,318	14,400	10,609	6,146	4,330
3.0	12,027	10,000	7,367	4,268	3,007
3.5	8,836	7.347	5,413	3,136	2,209
4.0	6,765	5,625	4,144	2,401	1,691
4.5	5,345	4,444	3,274	1,897	1,336
5.0	4,330	3,600	2,652	1,537	1,082
7.5	1,924	1,600	1,179	683	481
10.0	1,082	900	663	384	271
12.5	693	576	424	246	173
15.0	481	400	295	171	120
17.5	353	294	216	125	88
20.0	271	225	166	96	68
22.5	214	178	131	76	53
25.0	173	144	106	61	43
27.5	143	119	88	51	36
30.0	120	100	74	43	30

[a] *Example.* Find the minimum number of observations required to make a reliable statistical decision on a standard spinner's job of 12 sides at 40 ends down per 1000 spindle hours. The tolerance, %*T*, for this job is 12.5, and we would like to assume a 0.27% risk.

Procedure. Find %*T* value in column 1, and follow the horizontal line across the table to the column corresponding to the risk 0.27%. With a %*T* of 12.5 and 0.27% risk the minimum number of observations are 576.

Equation. $N = (100t/T)^2$. Thus, for $t = 3.0$ (corresponding to an 0.27% risk), and $T = 12.5\%$, $N = (100 \times 3/12.5)^2 = 576$.

cal decisions to be made regarding job loads. At this point, and thereafter, it is obvious that when the number of observed end breaks exceeds the job tolerance a reduction in the spinner's assignment (one frame side) should be considered. On the other hand, an observed end breakage below the tolerance indicates that the spinner's assignment might be increased by one side.

Values of *N*

Table 5-4 provides *N* values for a wide range of specifications and conditions for general use. To use this table, first determine the job tolerance required, $\%T$, and then follow the horizontal line across the table to the column corresponding to the risk considered allowable. The *N* value at the intersection represents the accumulated number of defects or attributes that must be observed before making a statistical decision regarding the job load.

For that sample size, the sampling error, $\%E$, is approximately equal to the job tolerance, $\%T$. After making an inspection test for the designated number of occurrences, *N*, calculate the observed average frequency and compare the results with the job standard tolerance.

Variation in Occurrence Rates

It has already been noted that industry tolerates a certain amount of variation, from minute to minute, hour to hour, and day to day. This allowance is based upon consideration of normally expected variations in raw stock and atmospheric conditions, as well as in such elements as maintenance and cleaning schedules, and a multitude of other processing factors that cannot be controlled at absolute precision, and are thus considered a part of practical operating conditions. These variables superimpose a longer term variation in quality and running performance over the short term fluctuations considered in the preceding discussion.

Actual experience indicates that over a period of several days such rates as ends down in roving and spinning and loom stops in weaving exceed the theoretical short term expectations by 50 to 100 per cent. This is graphically illustrated in Figure 5-5 and may be expressed by the following equation:

$$\sigma_T^2 = \sigma_w^2 + \sigma_b^2 \qquad (6)$$

The squared standard deviation, σ^2, is called the *variance*. The subscripts, in each case, refer to the type of variance; *T* equals total, w equals within short term periods, and b is equal to between short term periods. Together, the within short term and the between short term (but within long term) sources of variation comprise over-all or total variation.

The analysis of variations in a mill, by using variance components, was pioneered by L. H. C. Tippett of the British Cotton

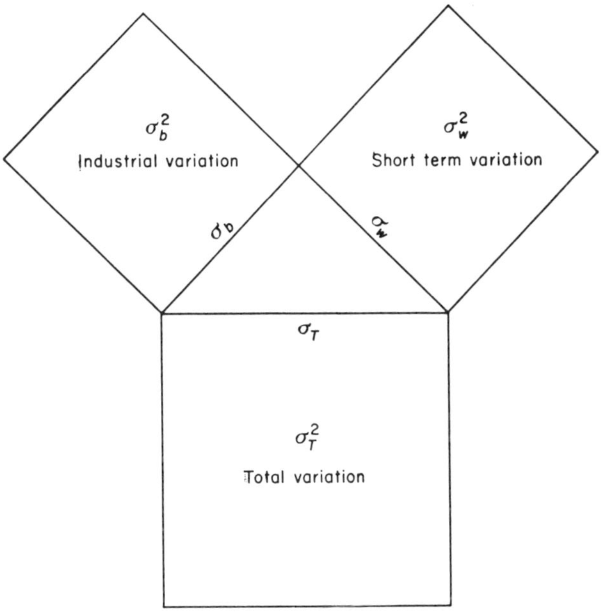

Figure 5-5. Total variance, σ_T^2, is composed of two elements, namely, short term variation (within short time periods) and variation between short terms (designated as industrial variation).

Industry Research Association (3), and subsequently an expanded method was developed in this country and has found wide use in American mills (1). A simplified method of finding σ_w, σ_b, and σ_T is given by Enrick (2), and covers various practical mill problems.

Proper application of statistical procedures requires that wherever possible σ_T be substituted for σ_w, and that sampling inspection be stratified to avoid decisions based on an isolated observation period. Usually, after σ_T and σ_w have been determined, the ratio $\sigma_T/\sigma_w = k$ can be used as follows to modify the previously given formula for N:

$$N_p = (100 \times t \times k/T)^2 \tag{7}$$

where N_p represents the practical sample size, considering long term fluctuations. N_p can also be found from Table 5-4, by multiplying N by k^2.

Stratified Sampling

Stratification of sampling and observation periods recognizes variation in quality level at various times and areas. Continuous sampling of one stratum invariably produces a biased sample. Data from different strata may be added to give an accumulated result which can then be interpreted.

Consideration should also be given to the composition of stratified samples, based on cost factors and requisite reliability. Suppose we wish to consider 1 hour inspection tests per stratum, as compared to 4 hour tests. Using V to denote the variation coefficient (the standard deviation expressed as a percentage of its average, as explained earlier in this chapter), as well as the subscripts w and b, and with W and B representing the relative costs per hour of 4 continuous hours and 4 single-hour stratified inspections, respectively, we obtain:

$$n = V_w\sqrt{B/W}/V_b \tag{8}$$

where n is the number of consecutive 1 hour tests per stratum. For example, if V_w is 5%, V_b is 7%, and the ratio B/W is 2, thus representing the inconvenience and extra effort required to change strata, we have:

$$n = 5\%\sqrt{2}/7\% = 1.01 \tag{9}$$

One hour inspection tests, randomly distributed among strata, will thus provide the most reliable sample at the least inspection costs.

References

1. N. L. Enrick, *Statistics as a Tool in Quality Management*, paper presented at the Annual Textile Division Meeting, American Society for Quality Control, Aug. 24, 1951.
2. N. L. Enrick, *Quality Control and Reliability*, 7th ed., Industrial Press, New York, 1977.
3. L. H. C. Tippett, *Shirley Inst. Mem.*, 13, 79-93 (1934).

CHAPTER 6

Work Sampling—Principles and Procedures

Introduction

Work sampling is a valuable measurement tool for the industrial engineer. It is based upon the fact that from an adequate sample of randomly spaced instantaneous or "snap" observations of an activity, conclusions can be drawn concerning the whole activity. Moreover, these conclusions can be arrived at with a known degree of confidence.

To illustrate the manner in which work sampling may be used, let us consider the conclusions that can be reached from 1000 random observations of a machine in the case where 750 observations showed the machine to be running and 250 observations showed it to be stopped or "down." The best conclusion that can be reached from these data is that during the period covered by the observations the machine was down 250 times out of 1000, or 25 per cent of the time. As will be shown later, the mathematical theory of work sampling permits us to make a further statement. We can say that there are only 3 chances in 1000 that the percentage down time of the machine during the period studied was outside the range of 25 ± 4.1 per cent, or from 20.9 to 29.1 per cent. If we are willing to accept a possibility of error greater than 3 chances in 1000, then we can conclude that the down time percentage fell within a narrower range. More observations would permit us to narrow the range without increasing our chances of being wrong.

Application

What, then, are the applications of work sampling? The five major areas in which work sampling is particularly applicable will be mentioned here:

1. Work sampling is useful in nonstandardized work where it is desired to establish the frequency and the time requirement of various operations. Direct measurement by means of time study is often impractical or uneconomical as compared with work sampling.

2. Work sampling is often the most economical means of assembling initial survey data. It can be used to establish the relative amounts of time invested in the various activities of a department.

3. Work sampling provides an effective means of measuring the degree of stability of the conditions upon which time study standards have been established. Continuing work sampling studies can provide a review of the elemental occurrence factors used for the establishment of standards.

4. Work sampling can often be carried out by a supervisor or by a member of the department involved. This is an economical method of study and also has the psychological advantage that the department involved is conducting its own study.

5. Work sampling can be used to obtain data on operations that would be affected by time study. For instance, on a machine-paced job where an operator has a choice between continuous visual product inspection and idle time, the operator would tend to give greater emphasis to inspection during a time study. Work sampling could be carried out in a manner to avoid this distortion.

Probability Concepts

Work sampling employs the concept of probability. To briefly examine this concept, consider some of the probabilities involved in rolling a die. The probability of rolling a 5 is $^{1}/_{6}$. The same probability exists for rolling each of the six numbers on the die. The sum of their six individual probabilities is $^6/_6$ or 1.00. This value, 1.00, always represents the total of the individual probabilities of mutually exclusive events.

In the case of our die the probability, p, of rolling a particular number would be $^1/_6$ (0.167 or 16.7 per cent). The probability of not rolling that number would be $1 - p$ or 1-0.167 which equals 0.833 or 83.3 per cent. The probability of *either* of two mutually exclusive events occurring is equal to the sum of their individual probabilities. Thus, the probability of rolling a 3 or a 5 is 0.167 + 0.167 or 0.334. The probability of rolling a 5 on each of two rolls is the product of the probability for each individual roll; therefore, the probability is 0.167 × 0.167 equal to 0.028. In general, the probability of an event occurring on n successive occasions is p^n. These concepts are further illustrated in Figure 6-1.

The lot consists of 5 marbles, 25% black and 75% white. The odds that 5 randomly drawn marbles will all be white are:

$$0.75 \times 0.75 \times 0.75 \times 0.75 \times 0.75 = 0.24$$

However, one black marble may occur in any one of these combinations, and thus, the odds for this occurrence would be:

●	○	○	○	○	=	0.25 × 0.75 × 0.75 × 0.75 × 0.75 = 0.08
○	●	○	○	○	=	0.75 × 0.25 × 0.75 × 0.75 × 0.75 = 0.08
○	○	●	○	○	=	0.75 × 0.75 × 0.25 × 0.75 × 0.75 = 0.08
○	○	○	●	○	=	0.75 × 0.75 × 0.75 × 0.25 × 0.75 = 0.08
○	○	○	○	●	=	0.75 × 0.75 × 0.75 × 0.75 × 0.25 = 0.08
						0.40

Therefore, the probability that a sample of 5 marbles will contain all whites *or* just one black is equal to

$$0.24 + 0.40 = 0.64$$

Figure 6-1. A further illustration of the probability law.

In work sampling we are interested in the probability of an event occurring a given number of times in a sample. For instance, what is the probability of the number 5 coming up four times in 10 rolls of the die? The determination of this probability is based upon an extension of the calculations just shown, by using formulas derived from the binomial theorem. The formula for the probability, in a sample of n values, of an exact number of occurrences, r, is as follows:

$$\text{Probability of } r \text{ occurrences} = \frac{p^r \times (1\text{-}p)^{n-r} \times n!}{r!\,(n\text{-}r)!} \qquad (1)$$

where $n!$ is read "n-factorial" and denotes $1 \times 2 \times 3 \times \ldots \times n$. For the illustration at hand (n equals 10, r is equal to 4, and p is 0.167) the solution is as follows:

1. The relation of the factorial expressions is equal to $n!$ divided by $r!(n\text{-}r)!$, or $10!/(4! \times 6!) = 210$.

2. Next, multiplying 210 by the remainder of the equation, $0.16^4 \times 0.833^6$, yields 0.0545 as the probability of four occurrences in a sample of 10.

There are then about 5 chances in 100 of having four 5's come up in 10 rolls of the die, If, for large values of n, we plot r on a horizontal scale and the probability of r occurrences on the vertical scale, we obtain a graph that approximates a normal curve. The normal curve, as discussed below, is of great practical value in work sampling.

The Normal Curve

The normal curve requires a considerable mathematical background for its derivation. However, an explanation of how

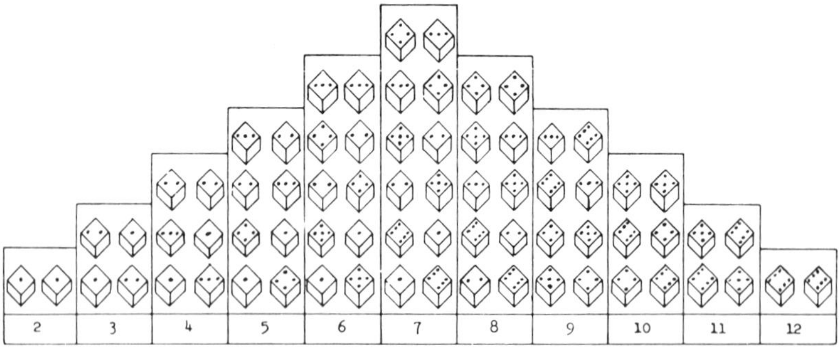

Figure 6-2. Distribution expected from 36 throws of a pair of dice. This distribution is a forerunner of the normal curve shown in Figure 6-3. (From N. L. Enrick, *Quality Control*, 4th ed., Industrial Press, New York, 1960. Reproduced by permission.)

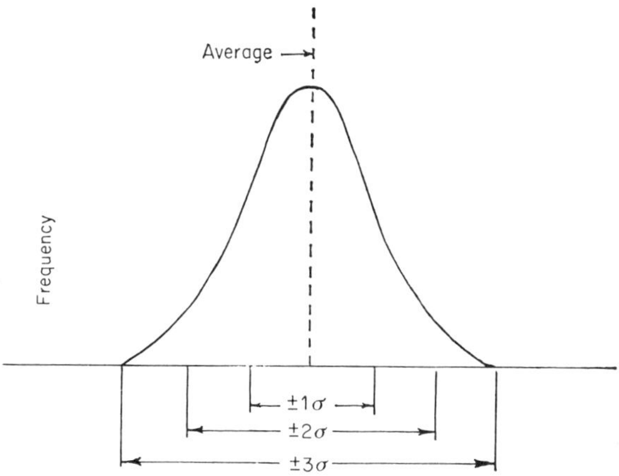

Figure 6-3. The normal curve showing limits of 1, 2, and 3 standard deviations, σ.

it arises has been given from the probability calculations in the preceding section. An intuitive understanding of the normal curve can be gained from a study of the throws of dice, as shown in Figure 6-2. Whereas individual distributions of dice throws will vary, the most likely result obtained from 36 throws is that presented by the diagram, where the throw of "snake eyes" (2) and "box cars" (12) are the least frequent, and 7 predominates. This distribution shapes itself after the fully normal curve pattern of Figure 6-3. Now, imagine that distributions were made from the throws of three dice, then four, and so on. Finally, the last and theoretically "infinite" distribution would assume the perfect normal outline.

The normal curve describes the distribution of sample values taken in a random manner from an unchanging mass of data. Work sampling studies supply samples of this type. Therefore, by referring to the normal curve, we can determine what the probability of the results of a work sampling will be, within given limits of the true values of the activity under study.

The normal curve can be described precisely in mathematical terms. It is defined by its average value and its dispersion. Since this curve is symmetrical, its average value is easily defined. Some method, however, is required to define the distribution about the mean. The most common measurement of distribution is the standard deviation, usually represented by σ, and defined as the square root of the sum of the squares of the deviations of the individual values.

The normal curve contains known percentages of the total values within distances measured in terms of the number of standard deviations from the mean, $\bar{X}$:

Limits	Approximate area under the curve, %
$\bar{X} \pm 1\sigma$	68
$\bar{X} \pm 2\sigma$	95
$\bar{X} \pm 3\sigma$	99.7

Thus, we are able to say that a sample value will fall within $\pm 1\sigma$ limit of the true value of the whole activity 680 times for every 1000 samples. The sample will be within $\pm 3\sigma$ limits of the true value of the whole activity 997 times for every 1000 samples. These are the most probable ratios for any number of samples and would be the correct ratios for an infinite number of samples from a normal curve.

Use of the Normal Curve

In a work sampling study use is made of our knowledge of the normal curve. In the study of machine down time the true value of the probability of down time, p', is estimated by means of the average value, p, of a sample of n observations. Equation (2) estimates the standard deviation of p for any given sample size, n:

$$\sigma_p = (p(1 - p)/n)^{1/2} \tag{2}$$

When p equals 0.25 and n is 1000

$$\sigma_p = (0.25 \times 0.75/1000)^{1/2} = 0.0137 \text{ or } 1.37\% \tag{3}$$

At a confidence level of 99.7 per cent (representing a risk of error of 3 in a 1000) the p of 0.25 lies within $\pm 3\sigma$, or $3 \times 1.37 = 4.1$

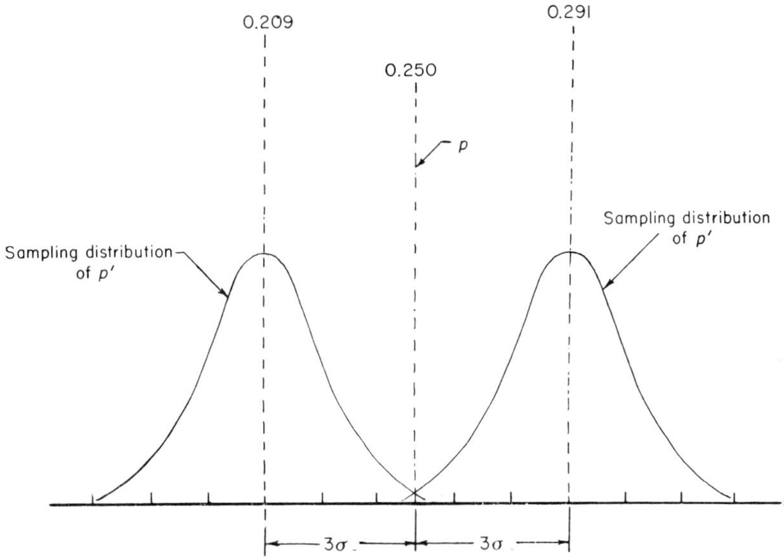

Figure 6-4. Sampling of an activity based on 1,000 observations. The range within which p' must fall is shown by $\pm 3\sigma$ (0.209 to 0.291), calculated from p for 99.7 per cent confidence limits.

per cent of the true value, p'. Figure 6-4 illustrates the basis of this prediction. Since we know that for confidence level of 99.7 per cent p falls somewhere under the normal curve for p' (within $\pm 3\sigma$ of p'), we also know that at the same confidence p' is within $\pm 3\sigma$, or in other words within ± 4.1 per cent of the value of p.

We have just introduced the term *confidence level*. This is an important concept, since in statistical sampling we are never able to state, with absolute certainty, the true value of an activity. We can state, however, that the true value lies within a certain range of values at a certain confidence level. The use of $\pm 2\sigma$ as the range, permits reporting the results with an accuracy of 95 per cent (i.e., the risk of error is 5 in 100). It would be rare in work sampling that a spread of less than $\pm 2\sigma$ be used. If the 95 per cent confidence level is not acceptable, the range may be extended to $\pm 3\sigma$ which increases the confidence level to 99.7 per cent (i.e., the risk of error is reduced to 3 in 1000). Extension of the range beyond $\pm 3\sigma$ on the asymptotic portion of the normal curve is so close to the horizontal axis that an increase in the range of standard deviations results in only a negligible increase in the confidence level.

Relation of the Number of Observations to the Standard Deviation

The standard deviation is inversely proportional to the number of observations. Referring back to equation (3) we observe that if 2000 observations had been made, then

$$\sigma_p = (0.25 \times 0.75/2000)^{1/2} = 0.00968 \qquad (4)$$

and $\pm 3\sigma = \pm 0.02904$ or ± 2.9 per cent, representing confidence limits of 25 ± 2.9 per cent = 22.1 to 27.9 per cent.

Thus, by increasing the number of observations, the range can be narrowed or the confidence limits improved.

Applying Work Sampling

We have now discussed the basic tools necessary for a work sampling study, provided that the study is of an activity that varies only as a result of statistical chance, and that the observations are made purely at random. The next section deals with checks that will assure fulfillment of the requirements.

First, let us assume that these conditions exist, and show by means of example (the job of a winder operator on a resin finishing range) the steps that must be covered. We are interested in determining the relative proportions of three conditions:

1. Machine down.
2. Machine running, work being performed.
3. Machine running, operator idle or performing inspection.

Table 6-1 presents the results of an initial series of 200 observations.

First, consider machine down time. The value of $p = 0.215$. Calculate p' within a range of ± 2.5 per cent and with a con-

TABLE 6-1

Category	No. of observations	Occurrence	
		Per cent	Fraction
Machine down	43	21.5	0.215
Machine running, operator working	64	32	0.320
Machine running, operator idle or inspecting	93	46.5	0.465
Total	200	100.0	1.000

fidence of 99.7 per cent. Therefore, since $\pm 3\sigma$ yields a range of ± 0.025, we must make sufficient additional observations to obtain a standard deviation of 0.025/3, or 0.0083. Next, by using a formula which gives σ as a function of n, we begin with equation (2):

$$\sigma_p = (p(1-p)/n)^{1/2} \tag{2}$$

Squaring and transposing,

$$\begin{aligned} n &= p\,(1-p)/\sigma_p^2 \\ &= 0.215 \times 0.785/0.0083^2 = 2450 \end{aligned} \tag{5}$$

An n of 2450 therefore represents the required number of observations, or more simply, the "sample size." A similar analysis will give the number of observations required for the other categories.

Machine running, operator working:

$$n = 0.32 \times 0.68/0.0083^2 = 3160$$

Machine running, operator idle or inspecting:

$$n = 0.465 \times 0.535/0.0083^2 - 3610$$

It will be necessary to make 3610 observations to satisfy all requirements. Assuming eight ranges, each observed at an average of once every half hour, then 28 shifts would be required. The data could be obtained by a member of the industrial engineering department in conjunction with similar studies of other machines. Alternative means would be to have the department supervisor or an employee make the observations.

The results of 4000 observations, grouped in units of 200, are given in Tables 6-2 and 6-3.

TABLE 6-2

Category	Occurrences: Per cent	Occurrences: Fraction
Machine down	21.3	0.213
Machine running		
Operator working	38.9	0.389
Operator idle or inspecting	39.8	0.398
Total	100.00	1.000

TABLE 6-3
Work Sampling Results on Finishing Ranges

Observation group no.	Machine down No.	Machine down Per cent	Machine running: Operator working No.	Machine running: Operator working Per cent	Machine running: Operator idle or inspecting No.	Machine running: Operator idle or inspecting Per cent
1	43	21.5	64	32	93	46.5
2	44	22	77	38.5	79	39.5
3	27	13.5	84	42	89	44.5
4	38	19	86	43	76	38
5	49	24.5	85	42.5	66	33
6	43	21.5	82	41	75	37.5
7	42	21	75	37.5	83	41.5
8	31	15.5	71	35	98	49
9	47	23.5	79	39.5	74	37
10	50	25	68	34	82	41
11	44	22	76	38	80	40
12	41	20.5	87	43.5	72	36
13	53	26.5	80	40	67	33.5
14	45	22.5	71	35.5	84	42
15	45	22.5	74	37	81	40.5
16	40	20	91	45.5	69	34.5
17	50	25	79	39.5	71	35.5
18	39	19.5	75	37.5	86	43
19	44	22	76	38	80	40
20	35	17.5	78	39	87	43.5
Total	850	21.3	1558	38.9	1592	39.8

Calculating 99.7 per cent confidence limits, the following is obtained for the machine down category:

$$\sigma_p = (0.213 \times 0.787/4000)^{1/2} = 0.0065$$

$$\pm 3\sigma = \pm 0.0195$$

$$0.213 \pm 0.0195 = 0.193 \text{ to } 0.223$$

Table 6-5 shows the error limits for all categories on the basis of 99.7 per cent confidence levels.

TABLE 6-4

Category	Range of occurrences		
	Low	Average	High
Machine down	19.3	21.3	23.3
Machine running			
Operator working	36.6	38.9	41.2
Operator idle or inspecting	37.5	39.8	42.1

For a 95 per cent confidence level, the convenient reference Table 6-5 may be used to obtain the required number of observations, n, corresponding to limits of ± 0.01 to ± 0.05.

Randomness

In the discussion of a work sampling study it has been assumed that the observations were made in a random manner. Randomness is necessary, since work sampling theory is based on this assumption. Randomness of observations requires that each interval of time throughout the period studied have equal probability of use for observation. We shall discuss here the means of developing randomness, and the means of testing to determine whether randomness has been achieved.

The most convenient means of establishing random observation times is through use of a set of random sampling numbers, as provided in Table 6-6,where the numbers appear in completely random form. As an example, if it is desired to make 10 observations during the period of an hour, enter the table at a random

TABLE 6-5
Sample Sizes Required for Various Limits of Error (95 Per Cent Confidence Level)

p	Sample size required for confidence limits at 95%					*p*
	±0.01	±0.02	±0.03	±0.04	±0.05	
0.01	396[a]	100[a]	44[a]	25[a]	16[a]	0.99
0.02	784	196[a]	88[a]	49[a]	32[a]	0.98
0.03	1,163	292	130[a]	73[a]	47[a]	0.97
0.04	1,535	384	171	96[a]	62[a]	0.96
0.05	1,900	475	212	119	76[a]	0.95
0.06	2,260	565	252	142	92[a]	0.94
0.07	2,604	654	290	163	102	0.93
0.08	2,945	738	328	184	118	0.92
0.09	3,278	820	364	205	131	0.91
0.10	3,600	900	400	225	144	0.90
0.11	3,918	980	435	245	157	0.89
0.12	4,224	1,055	470	264	169	0.88
0.13	4,520	1,130	504	282	181	0.87
0.14	4,820	1,210	535	302	193	0.86
0.15	5,100	1,275	568	318	205	0.85
0.16	5,380	1,350	600	337	216	0.84
0.17	5,650	1,415	628	353	226	0.83
0.18	5,900	1,475	656	369	236	0.82
0.19	6,160	1,545	685	385	246	0.81
0.20	6,410	1,605	715	400	256	0.80
0.21	6,640	1,660	740	415	266	0.79
0.22	6,870	1,720	765	430	275	0.78
0.23	7,100	1,780	790	444	284	0.77
0.24	7,300	1,830	815	456	292	0.76
0.25	7,500	1,880	835	470	300	0.75
0.26	7,690	1,925	855	481	308	0.74
0.27	7,885	1,970	875	493	316	0.73
0.28	8,065	2,015	895	504	323	0.72

position. Then, read along the table diagonally, vertically, or horizontally, recording the first 10 numbers from 01 through 60. We would then arrive at the observation times found as indicated in Table 6-7.

If the observer is making a series of observations during trips through the plant, random selection of times for starting each trip would usually be satisfactory.

Testing for randomness is done with a statistical control chart known as a *c*-chart. In control charting for quality control, $\bar{c}$ denotes the average number of defects per unit; and corre-

TABLE 6-5 (*continued*)

p	Sample size required for confidence limits at 95%					p
	±0.01	±0.02	±0.03	±0.04	±0.05	
0.29	8,240	2,060	915	515	330	0.71
0.30	8,400	2,100	935	526	337	0.70
0.31	8,555	2,140	950	535	343	0.69
0.32	8,705	2,175	965	545	349	0.68
0.33	8,840	2,210	985	553	354	0.67
0.34	8,975	2,245	1,000	561	360	0.66
0.35	9,100	2,275	1,010	569	365	0.65
0.36	9,220	2,305	1,025	576	369	0.64
0.37	9,325	2,330	1,035	583	373	0.63
0.38	9,425	2,355	1,045	589	377	0.62
0.39	9,515	2,380	1,055	595	381	0.61
0.40	9,600	2,400	1,065	600	384	0.60
0.41	9,675	2,420	1,075	605	387	0.59
0.42	9,745	2,435	1,085	609	390	0.58
0.43	9,805	2,450	1,090	613	392	0.57
0.44	9,855	2,465	1,095	616	395	0.56
0.45	9,900	2,475	1,100	619	397	0.55
0.46	9,935	2,485	1,105	621	398	0.54
0.47	9,965	2,490	1,110	623	399	0.53
0.48	9,985	2,495	1,110	624	400	0.52
0.49	9,995	2,500	1,115	625	400	0.51
0.50	10,000	2,500	1,115	625	400	0.50

Source: R. E. Heiland and W. J. Richardson, *Work Sampling*, McGraw-Hill, New York, 1957, p. 234.

[a] Since, as a rule of thumb, np should equal 5 or more, these values should be increased to meet this criterion. For example, for p — 0.03, n should be increased from 130 to 167, so that 0.03(167) = 5.

spondingly in work sampling, $\bar{c}$ can be used to denote the average number of observations per time period. For this chart, σ_c is equal to $\sqrt{\bar{c}}$. This formula is derived from the Poisson distribution, which approximates the binomial distribution when p is relatively small (less than 10 per cent) and n is larger than 50. The $\pm 3\sigma$ limits are

$$\text{confidence limits} = \bar{c} \pm 3\sqrt{\bar{c}}$$

Figure 6-5 shows the application of a c-chart to the study of the range winders. Examine the stability shown when the observations are grouped in half-hour intervals each day. In this case,

TABLE 6-6
Random Sampling Numbers[a]

10	05	65	97	10	00	68	73	30	31	45	19	17	02	16	39	17	50	29	67
02	85	35	47	95	87	37	52	56	07	07	85	14	31	71	88	71	16	50	85
83	27	95	62	63	73	57	03	99	36	24	18	63	02	98	00	26	73	25	24
62	44	46	17	19	68	69	07	33	83	56	16	00	25	20	84	40	22	06	92
50	34	36	69	05	49	79	39	37	17	23	25	08	91	90	24	48	89	35	49
22	82	64	93	38	71	42	13	88	32	60	65	58	94	22	95	09	13	36	46
11	19	80	19	54	21	10	41	33	13	69	05	43	71	01	47	70	02	43	04
38	18	82	51	45	08	36	74	38	74	56	48	39	10	62	89	53	13	18	14
83	55	50	22	49	72	73	57	55	17	06	06	06	87	68	98	09	01	96	34
54	74	74	46	88	17	29	24	40	59	53	45	83	40	80	68	83	07	93	33
49	07	95	12	80	32	73	63	02	53	95	86	94	09	45	25	88	10	76	16
60	67	26	60	73	78	52	59	32	19	25	32	92	57	38	35	39	00	72	49
29	29	72	83	68	09	95	90	69	01	15	46	30	65	12	55	61	03	40	35
65	73	13	73	74	35	24	85	40	06	97	42	22	19	07	26	94	57	66	80
86	65	60	69	73	96	20	05	52	85	94	81	24	50	16	96	30	72	25	76
00	93	99	36	04	50	06	96	74	57	81	23	93	49	63	90	98	26	63	59
40	08	58	79	34	65	55	64	86	85	51	52	33	29	18	60	43	81	80	29
19	64	24	74	97	64	84	37	32	06	28	70	50	28	10	95	00	39	00	56
21	28	10	07	75	45	25	92	26	09	88	33	89	08	54	61	70	55	10	45
64	79	47	34	02	47	34	02	47	36	27	68	29	77	08	90	04	28	81	18
26	50	64	58	98	27	68	64	47	65	82	76	36	70	09	22	31	20	41	03
53	59	36	53	76	51	13	99	53	68	35	70	05	75	64	63	19	85	03	01
43	85	39	15	21	21	97	50	40	00	44	51	94	96	90	81	58	42	01	95
04	71	52	62	92	74	21	10	43	43	74	47	40	94	95	68	76	54	68	94
57	16	52	55	27	27	73	07	18	74	24	46	78	29	33	16	48	29	10	92
29	52	51	85	86	64	55	65	34	79	58	08	63	49	93	09	14	60	85	82
46	23	81	57	74	96	06	50	04	36	99	93	16	50	24	02	50	96	62	40
68	81	94	85	52	05	20	96	73	69	60	65	07	19	22	09	62	26	26	72
06	42	97	54	81	17	67	41	74	68	50	95	43	24	70	77	54	97	14	47
35	98	34	91	25	62	37	96	45	75	30	35	29	64	77	91	33	99	45	83
58	87	29	39	26	79	67	06	16	27	20	79	95	31	30	44	60	49	42	03
76	81	67	77	49	88	89	50	72	15	72	97	89	79	17	82	39	45	82	98
78	90	64	51	69	25	58	24	37	86	98	94	75	97	90	13	69	47	89	69
27	97	97	21	28	92	20	79	05	56	18	55	04	99	53	21	98	52	60	84
14	68	53	35	31	53	60	30	57	87	36	62	01	45	48	83	86	64	35	81
58	88	28	02	45	49	08	80	21	06	55	72	66	12	88	03	61	53	77	15
39	13	90	60	97	27	61	48	93	10	05	99	18	59	48	65	59	20	70	53
14	52	12	81	25	81	36	00	54	77	30	01	88	25	44	49	30	18	59	37
81	90	03	17	23	05	74	63	31	96	57	64	79	48	72	77	66	28	70	19
83	83	44	70	11	25	59	56	28	44	17	20	83	66	71	24	14	42	55	42

[a] N. L. Enrick, *Statistics Course Notes*, 1960.

$\bar{c}$ is equal to 4000 observations per 48 periods or 83.3. The upper control limit becomes $83.3 + 3\sqrt{83.3}$ or 110.6, whereas the lower control limit is 56.0. The plot of values pertaining to this study shows stability. No points fall beyond the $\pm 3\sigma$ limits. The points form a roughly normal distribution between the $\pm 3\sigma$

TABLE 6-7
Determination of Observation Times from Table of Random Numbers

Values read from table of random numbers[a]	Observation times from 8 to 9 A.M. (corresponding to random numbers)[b]	Observation times arranged in scheduled sequence[c]
19	8:19	8:02
26	8:26	8:17
45	8:45	8:19
39	8:39	8:23
17	8:17	8:26
47	8:47	8:29
38	8:38	8:38
23	8:23	8:39
02	8:02	8:45
29	8:29	8:47

[a] Since the random numbers are to correspond to minutes, record only the first two digits from the Table 6-6. Also, omit any numbers in excess of 60.

[b] For this example, with observations between 8 and 9 A.M., consider the units of the values in the first column to be minutes, and add 8.

[c] The data have been arranged in orderly time sequence for the purpose of scheduling actual observations.

limits, and therefore, it can be concluded that the observations were obtained in a random manner.

Control Charts

Control charts for p can be used to determine whether a change is occurring in the value of p', thus providing a valuable means of detecting change through the use of continuing studies. The chart represents a plot of successive values of p. A center line is drawn representing $\bar{p}$. Parallel lines are drawn at $\pm 1\sigma$, $\pm 2\sigma$, and $\pm 3\sigma$. The plot of a value outside the $\pm 3\sigma$ limits would be expected to have an occurrence of 0.3 per cent. With confidence limits set at 99.7 per cent, a point on the chart outside the $\pm 3\sigma$ lines would indicate that a significant change has occurred.

With the control chart it is not necessary to wait until a point falls beyond the $\pm 3\sigma$ limits to conclude that a change has taken place. The *theory of runs* can also be utilized to detect

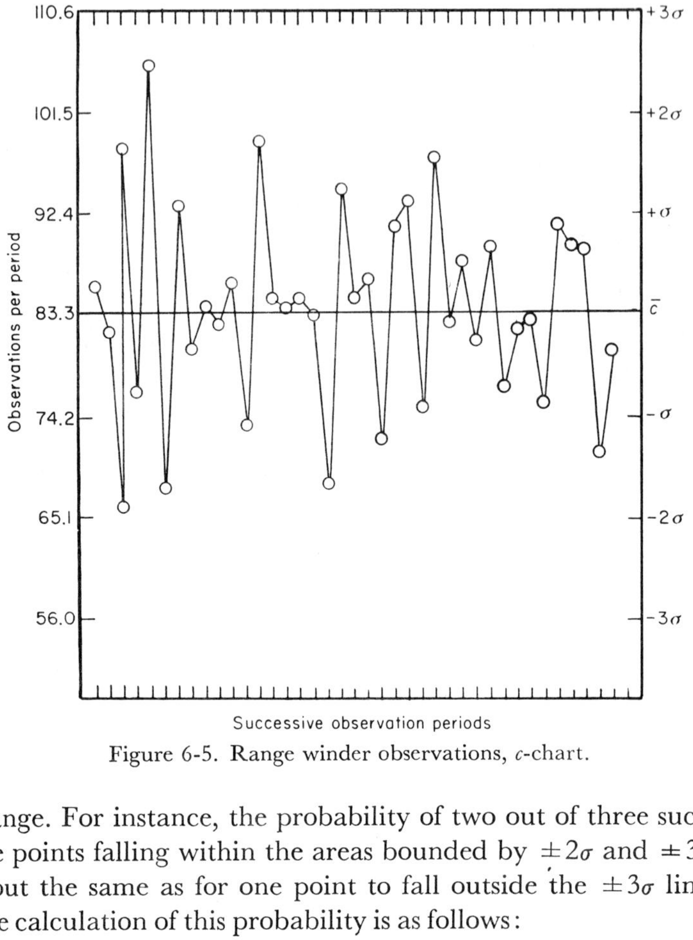

Figure 6-5. Range winder observations, *c*-chart.

change. For instance, the probability of two out of three successive points falling within the areas bounded by $\pm 2\sigma$ and $\pm 3\sigma$ is about the same as for one point to fall outside the $\pm 3\sigma$ limits. The calculation of this probability is as follows:

$$\text{Probability of occurrence of exact number of prescribed events} = \frac{p^r \times (1\text{-}p)^{n-r} \times n!}{r!(n-r)!} \tag{6}$$

where n is the number of points in the run; r is the number of points within the prescribed limits; and p is the probability of each point being within the prescribed limits. When p is equal to 0.043 (the per cent of area of a normal curve falling between 2σ and 3σ limits), n is equal to 3, and r equals 2, the probability of a run is:

$$0.043^2 \times (1 - 0.043) \times 3!/2! = 0\ 00532$$

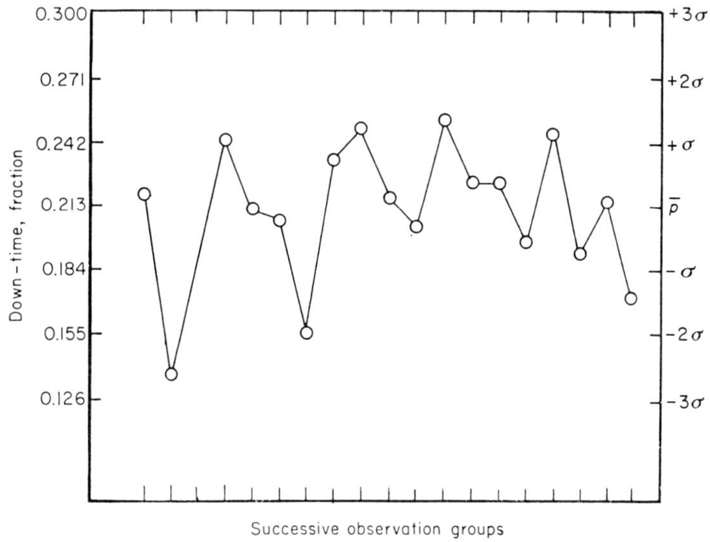

Figure 6-6. Range winder study. The machine down-time is given by observation groups and is expressed as a decimal fraction, and is plotted on a *p*-chart.

This value is the probability of two out of three points falling within the boundaries of $\pm 2\sigma$ and $\pm 3\sigma$. This probability of 0.0053 is of the same approximate order of magnitude as the 0.003 probability of one point falling outside $\pm 3\sigma$ limits.

Other runs of low probability can be calculated in a similar manner. A run in which four out of five points fall between $\pm 1\sigma$ and $\pm 3\sigma$ would likewise indicate that a change has occurred.

Thus, the control chart is a sensitive device for telling when the probability of change exceeds the allowed confidence limits. The discussion above has considered that we wish to be 99.7 per cent confident that change has occurred. If we wish to be alerted when the probability that change has occurred is only 95 per cent, we would use $\pm 2\sigma$ as the outside limits of our chart. Runs of the same probability can be calculated. In this manner both favorable and unfavorable trends can be detected.

The control chart also tells other facts concerning observed data. For example, the plotted points may not show a tendency to a normal distribution or they may be changing (cyclic or

TABLE 6-8

Estimated Ends Down Rate from Observed Percentage of Idle (Broken End) Spindles[a]

Idle spindles, %	Spinner's cycle time, minutes																
	15	20	23	25	26	27	28	29	30	31	32	33	34	35	36	38	40
0.1	8	6	5	5	5	4	4	4	4	4	4	4	4	3	3	3	3
0.2	16	12	10	10	9	9	9	8	8	8	8	7	7	7	7	6	6
0.3	24	18	16	14	14	13	13	12	12	12	11	11	11	10	10	9	9
0.4	32	24	21	19	18	18	17	17	16	15	15	15	14	14	13	13	12
0.5	40	30	26	24	23	22	21	21	20	19	19	18	18	17	17	16	15
0.6	48	36	31	29	28	27	26	25	24	23	23	22	21	21	20	19	18
0.7	56	42	37	34	32	31	30	29	28	27	26	25	25	24	23	22	21
0.8	64	48	42	38	37	36	34	33	32	31	30	29	28	27	27	25	24
0.9	72	54	47	43	42	40	39	37	36	35	34	33	32	31	30	28	27
1.0	80	60	52	48	46	44	43	41	40	39	38	36	35	34	33	32	30
1.1	88	66	57	53	51	49	47	46	44	43	41	40	39	38	37	35	33
1.2	96	72	63	58	55	53	51	50	48	46	45	44	42	41	40	38	36
1.3	104	78	68	62	60	58	56	54	52	50	49	47	46	45	43	41	39
1.4	112	84	73	67	65	62	60	59	56	54	52	51	49	48	47	44	42
1.5	120	90	78	72	69	67	64	62	60	58	56	55	53	51	50	47	45
1.6	128	96	83	77	74	71	69	66	64	62	60	58	56	55	53	51	48
1.7	136	102	89	82	78	76	73	70	68	66	64	62	60	58	57	54	51
1.8	144	108	94	86	83	79	77	74	72	70	68	65	64	62	60	57	54
1.9	152	114	99	91	88	84	81	79	76	74	71	69	67	65	63	60	57
2.0	160	120	104	96	92	89	86	83	80	77	75	73	71	69	67	63	60
2.2	176	132	115	106	102	98	94	91	88	85	82	80	78	75	73	69	66
2.4	192	144	125	115	111	107	103	99	96	93	90	87	85	82	80	76	72
2.6	208	156	136	125	120	116	111	108	104	101	98	95	92	89	87	82	78
2.8	224	168	146	134	129	124	120	116	112	108	105	102	99	96	93	88	84
3.0	240	180	157	144	138	133	129	124	120	116	112	109	106	103	100	95	90

Source: N. L. Enrick, paper presented at the Southern Textile Methods and Standards Association, March 22, 1956.

[a] Ends down per 1000 spindle hours = (1200 × idle spindles, %)/spinner's cycle time, minutes.

some other manner). This could either be inherent in the process, or it could represent inadequate training or supervisory control. Figure 6-6 shows a control chart for p that uses data from the study of the range winder. The chart indicates that stability existed during the period covered by the study.

Care must be employed not to apply work sampling to a work situation where stability does not exist. However, when applied properly, work sampling can be an effective means of measurement.

Patrol Walk-Through

A special application of work sampling, used widely in roving and spinning, is the determination of ends down by means of frequent "walk-through" checks. If the checker selects his path at random, then any end down he encounters would, on the average, have existed for a time equal to half the spinner's cycle. Therefore, by noting the number of nonproductive or idle spindles, a good indication of the number of ends down per 1000 spindle hours is obtained.

ILLUSTRATIVE EXAMPLE 6-1
Relationship Between Ends Down and Idle Spindles

With use of the time-saving method of patrol walk through, which yields data in "percent idle spindles," the formula below permits conversion to the conventional "ends down per 1000 spindle hours."

Definitions

E – ends down per 1000 spindle hours
I = idle spindles percentage
W = waiting time in minutes until spinner encounters and pieces broken end, which is usually one-half the cycle time.

Formulas for I and E

$$I = 100(W/60 \times E/1000) = WE/600$$

where 60 converts minutes to hours and 1000 converts 1000 hours to single hour. Solving for E, it is found that:

$$E = 600I/W$$

Example

Assume 0.6% idle spindles and 12 minutes for W, then $E = 600 \times 0.6/12 = 30$ ends down per 1000 spindle hours.

Extension of Formula

If instead of ends down per 1000 spindle hours, the terms ends down per hank or ends down per pound produced are preferred, the conversion formula is:

Ends down per hank

$$= \frac{E \times \text{inches/yard} \times \text{yards/hank} \times \text{contraction allowance}}{1000 \times \text{minutes/hour} \times \text{front roll rpm} \times \text{diameter, inches} \times \pi}$$

For the above example, with use of a front roll speed of 140 rpm, 1 inch diameter, and 5% contraction, it follows that:

$$\text{Ends down per hank} = \frac{30 \times 36 \times 840 \times 1.05}{1000 \times 60 \times 140 \times 3.142} = 0.036$$

Since this example used a 40's yarn, the ends down per pound are found from:

$$\text{Ends down per pound} = 0.036 \times 40 = 1.44$$

Source: N. L. Enrick, paper presented at the Southern Textile Methods and Standards Association, March 22, 1956.

Based on these principles, the relationships worked out in Illustrative Example 6-1 are derived. These in turn may be presented in the form of a convenient summary (Table 6-8). From Table 6-8 the relation between the percentage of idle spindles observed during work sampling may be related to the spinner's cycle time and to the estimated rate of ends down per 1000 spindle hours. Thus, if the patrol walk-through on one spinner's assignment has found 1 per cent idle spindles, and the spinner's standard cycle is 30 minutes, then the rate of ends down is 40 per 1000 spindle hours. In actual practice, however, an adjustment of the relationship in terms of individual mill conditions is necessary. These adjustments will usually be small, and the table will certainly serve as a good starting point for relating the per cent of idle spindles (found from patrol walk-through) against the more conventional ends down per 1000 spindle hours.

CHAPTER 7

Work Sampling—Case History*

Introduction—The Problem

A carpet manufacturer was engaged in a program to improve the plant's production control. Several objectives of this program could be met by improving office and planning procedures. The ultimate improvements, however, in addition to resulting in a better inventory situation, should also benefit shop operations. In order to have a complete appraisal of the progress of the entire program, some analysis of shop activity was necessary. It was decided to take a work sampling study, aimed particularly at measuring the time that could be saved by better production control.

Since capital investment in looms was quite considerable, any down time resulting from lack of yarn was expensive. The replacement cycle for the yarn was from six to eight weeks. Furthermore, good production control was necessary in order to keep inventory low, and service at the level demanded by the trade. Therefore, a loom standing idle for lack of yarn represented a matter of some concern to management. Over the year preceding the study, production control performance had been steadily worsening. This was not the fault of the mill, but the mill was being penalized through short runs, waits for yarn, and irregular weave schedules. Also every time a beam or creel had to be cut or changed, production was lost and the cost of direct labor was increased. Therefore, as the program of improvement proceeded in the office, the effectiveness of operation of the mill should also improve. The work sampling study was undertaken

* This case history has been adapted from R. E. Heiland and W. J. Richardson, *Work Sampling*, McGraw-Hill, New York 1957. (Reproduced by permission.)

to provide measurement of this improvement, and also to find how much improvement could be expected.

In addition to measuring the effects of production control on the shop, management wanted to check certain allowances given, and to establish a bench mark for the particular level of operation. The line supervisors were told to keep a careful record of unusual occurrences, to help establish this level.

One of the reasons for production control improvement was to permit better scheduling of personnel. Also, since the market was seasonal, it was possible that a different policy might be necessary at different times of the year; that is, that shorter runs of different types of carpet would be made at slack times. It was hoped that the study would give information which would aid management in formulating this policy.

Why Work Sampling Was Used

Work sampling seemed an obvious tool to obtain the type of information desired. Basically, information was desired concerning a condition which definitely was not scheduled, namely, running out of raw material. Furthermore, it was desired to observe the looms, and not the personnel. While some data on the personnel would of course be collected—since people run looms—loom utilization was what management was after. The system of timekeeping in effect made it quite difficult to obtain the needed information simply by examining time cards of the weavers.

To generalize, what was wanted was an over-all measurement of what was very definitely an undesirable nonstandard shop condition. To look ahead, management planned a program which it was hoped would result in change for the better. Therefore, whatever measurement technique was chosen should also have the characteristic that future comparisons could be made easily and consistently. In such situations, work sampling has proved to be quite effective.

Selection of Categories

In selecting the categories, the main objective was to separate out those delays which could be inferred to be a result of production control performance. In addition, the working categories were selected in such a manner that they could be checked against existing incentive rates. The categories use terminology of the weave shed, but require little further explanation. A "runout" means that the creel or beam is empty and that a new one must be tied into the loom. A "cutout" means that there is still yarn on the beam or creel, but that a change of color is wanted badly enough to warrant the expense of tieing a new beam or creel in to the loom. Runouts occur normally; cutouts reflect poor production control and are extremely expensive. The categories selected were as follows:

1. Running time.
2. Hand time in weave rate.
3. Beam change—runout.
4. Beam change—cutout.
5. Weaver absent.
6. Change creel—runout.
7. Change creel—cutout.
8. No pile yarn or pile beam.
9. No binder or stuffer beam.
10. No orders or not scheduled (loom idle).
11. No manpower.
12. Machine repair—major.
13. Wait for decision (production control).
14. Samples—production looms only.
15. Loom repair—done by fixer (mechanic).
16. All others.
17. Change overloom (to different pattern).

Collection of Observations

An observation sheet of a very simple design was used (Figure 7-1). From the observations made on this sheet, a series of tabulating cards were punched. This punching was quite simple, and made statistical analysis of the results a much more rapid procedure. Five rounds of observations were made each

Date 9-21-59

Time 1:48

Loom	Element	Loom	Element	Loom	Element	Loom	Element
333-1	1	310-1	1				
332-1	2	303-1	2	402-9	1	375-3	12
331-1	13	309-1	1	450-9	1	376-3	11
330-1	10	304-1	2	417-8	10		
329-1	1	308-1	11	403-9	16	424-8	1
328-1	1	305-1	1	416-8	1	423-8	15
327-1	1	307-1	2	419-8	6		
326-1	16	306-1	1	415-8	1	601-6	10
325-1	10			418-8	1	600-6	10
324-1	1	279-1	16				
323-1	3	280-1	1	500-4	8	628-7	7
322-1	15	281-1	11	550-5	10		

Figure 7-1. Loom utilization study observation sheet.

day, on the day shift only. It was decided that two weeks' observation would be sufficient. By that time almost seven thousand observations would have been obtained, which would give a level of reliability sufficient for the study's objective. The plant was working five and a half days per week.

There were no special precautions necessary in collecting the data. The observers were experienced men, who had done work sampling before. The employees were informed of the study's objectives, and of the fact that it would be taken.

Results

The results of the study are shown in Table 7-1. Category 10, No Orders or Not Scheduled (Loom Idle), was designated category B, and carried separately in the summation of results. This was done in order to verify certain allowances which were given as a percentage of scheduled loom operation. Also, by presenting results in this fashion, it was felt that available loom time could be obtained directly.

TABLE 7-1
Summary of Results of Work Sampling Study of Loom Activity Conducted by Industrial Engineering Department

	Occurrence, %				
Element	12/4 Velvet	14/4, 24/4 Velvet	3/4, 4/4, 4/4 Wilton	14/4, 12/4 Wilton	2/4 Wilton
1. Running time	61.3	52.2	32.3	54 1	52.3
2. Hand time in weave rate	10.6	15.0	8.9	11.1	14.3
3. Beam change—runout	7.6	9.6	2.3	2.2	4.5
4. Beam change—cutout	0.2		0.1		0.2
5. Weaver absent	3.9	4.9	6.9	4.8	2.1
6. Change creel—runout			0.7	2.6	2.8
7. Change creel—cutout			2.0	6.6	2.9
8. No pile yarn or pile beam	4.8	6.1	15.3	6.8	10.0
9. No binder or stuffer beam	0.3	0.7	0.4		
11. No manpower	5.7	2.2	0.5		0.2
12. Machine repair—major	0.9	1.9	1.8	2.2	1.0
13. Wait for decision	0.4		5.0		
14. Samples—production looms only	0.2		12.6		
15. Loom repair—fixer	0.8	2.6	3.2	3.5	3.5
16. All others	3.3	3.5	4.9	5.5	4.9
17. Change over loom		1.3	3.1	0.6	1.3
	100.0	100.0	100.0	100.0	100.0
A. No observations—orders available	2458	974	812	316	1217
B. No observations—no orders or not scheduled	342	1	488	184	58
C. No orders or not scheduled, % of *A*	13.9		60.1	58.2	4.8
Total number of observations	2800	975	1300	500	1275

The results were broken down by loom width, which was the scheduling breakdown most convenient in interpreting performance. Also, type of loom was considered. The significant values were as follows:

1. No Pile Yarn or Pile Beam, category 8, represented a failure of production control to have the proper raw material on hand. This category represented the possible area for direct improvement through better production control. It should have been almost zero.

2. Creel Cutout, category 7, is very undesirable. Supplementary records kept during the study showed that this too was caused by poor scheduling.

3. The 12/4 Velvet looms were working three shifts, and still could not meet customer demand. At the same time, category B (category 10 in original list), No Orders or Not Scheduled, was 13.9 per cent of the total. Since orders were available, it was obvious that scheduling was at fault.

4. Change Over Loom (category 17) was quite satisfactory. This is worthy of note, since the foremen had estimated that this activity took as much as 10 per cent of the loom time. This reflected favorably on production control. The inference was that yarn control was inadequate, but parts of loom scheduling of pattern change were satisfactory.

5. To take the 12/4 Velvet looms as an example, category 11 (No Manpower) represented a significant loss. What had happened was that the foreman hesitated to call in weavers because there might not be yarn for the loom, and thus wages would be lost.

6. For the most critical looms, possible areas for savings through better production control were considerable. Lumping together categories 4, 8, 9, 11, and 13, for the 12/4 looms, a total of 11.4 per cent is reached. This represents the worst situation, but of course this total is significant for all looms. In fairness, it must be added that only the velvet looms were really critically overloaded with customer orders.

7. In general, the time standards proved out. Results of the study agreed closely with most of the running time and service categories in the study.

The real worth of this study was not only that it measured the current operation, but also that as changes in procedure were instituted in the production control office, the effect of these changes on shop operation could be measured. It was planned to take check studies at intervals of every three months. In general, management could then use the control chart principles to measure change. Too often only subjective evaluation of over-all activity is available.

CHAPTER 8

Methods Time Measurement

Introduction

Methods time measurement (MTM) is a procedure specifically designed to analyze the basic motions comprising any manual operation. To each motion a predetermined time standard is assigned, the exact time value depending on the nature of the motion and the conditions under which it is performed. MTM studies are based on the following well established facts:

1. Any manual operation is made up of distinct and recognizable basic motions.
2. Each basic motion has a constant time value at the normal performance level.
3. The time values for all basic motions have been established on the basis of thorough research. The time values are MTM data.

Thus, MTM is basically a tool of methods analysis, yielding answers in terms of performance time.

Uses of MTM

The principal uses of MTM are described briefly in the following summary listing:

1. As a basis for developing better methods.
   a. Developing better methods prior to production of a new product.
   b. Improving existing methods.
   c. Guiding product design.
   d. Developing effective tool design.
   e. Selecting effective equipment.

2. Establishment of production or wage incentive standards.
   a. Establishing time standards.
   b. Developing time formulas and other standard data.
   c. Estimating costs.
3. Other.
   a. Training supervisors to become method conscious.
   b. Settling grievances in connection with production standards.
   c. Research, particularly in connection with methods, learning time, and performance rating.

Development of MTM Data

MTM data were derived in 1946 as a result of the detailed analysis of motion picture films covering various operations. At the same time, all the conditions surrounding the operation were recorded. During each film the operator under study was rated for skill and effort.

The resultant standard time for each motion is recorded on the MTM data cards (see Tables 8-1 through 8-10*). The unit of time used on the data card is 0.00001 hour which is called a *time measurement unit* (TMU). The MTM tables on the data card therefore show the required number of TMU's for an operator of average skill and effort, working under average conditions, performing the motion under study.

Properly applied, the data will give results of greater accuracy than those of the conventional time study procedure. The MTM analyst, however, must be properly trained in the recognition and classification of motions employed. Under no circumstances should these data be used by anyone other than a trained MTM analyst.

* Reproduced by permission of the MTM Association for Standards and Research, Ann Arbor, Michigan.

TABLE 8-1
Reach, R

Distance Moved Inches	Time TMU A	Time TMU B	Time TMU C or D	Time TMU E	Hand In Motion A	Hand In Motion B
¾ or less	2.0	2.0	2.0	2.0	1.6	1.6
1	2.5	2.5	3.6	2.4	2.3	2.3
2	4.0	4.0	5.9	3.8	3.5	2.7
3	5.3	5.3	7.3	5.3	4.5	3.6
4	6.1	6.4	8.4	6.8	4.9	4.3
5	6.5	7.8	9.4	7.4	5.3	5.0
6	7.0	8.6	10.1	8.0	5.7	5.7
7	7.4	9.3	10.8	8.7	6.1	6.5
8	7.9	10.1	11.5	9.3	6.5	7.2
9	8.3	10.8	12.2	9.9	6.9	7.9
10	8.7	11.5	12.9	10.5	7.3	8.6
12	9.6	12.9	14.2	11.8	8.1	10.1
14	10.5	14.4	15.6	13.0	8.9	11.5
16	11.4	15.8	17.0	14.2	9.7	12.9
18	12.3	17.2	18.4	15.5	10.5	14.4
20	13.1	18.6	19.8	16.7	11.3	15.8
22	14.0	20.1	21.2	18.0	12.1	17.3
24	14.9	21.5	22.5	19.2	12.9	18.8
26	15.8	22.9	23.9	20.4	13.7	20.2
28	16.7	24.4	25.3	21.7	14.5	21.7
30	17.5	25.8	26.7	22.9	15.3	23.2

CASE AND DESCRIPTION
A Reach to object in fixed location, or to object in other hand or on which other hand rests.
B Reach to single object in location which may vary slightly from cycle to cycle.
C Reach to object jumbled with other objects in a group so that search and select occur.
D Reach to a very small object or where accurate grasp is required.
E Reach to indefinite location to get hand in position for body balance or next motion or out of way.

TABLE 8-2
Move, M

Distance Moved Inches	Time TMU A	Time TMU B	Time TMU C	Hand in Motion B
¾ or less	2.0	2.0	2.0	1.7
1	2.5	2.9	3.4	2.3
2	3.6	4.6	5.2	2.9
3	4.9	5.7	6.7	3.6
4	6.1	6.9	8.0	4.3
5	7.3	8.0	9.2	5.0
6	8.1	8.9	10.3	5.7
7	8.9	9.7	11.1	6.5
8	9.7	10.6	11.8	7.2
9	10.5	11.5	12.7	7.9
10	11.3	12.2	13.5	8.6
12	12.9	13.4	15.2	10.0
14	14.4	14.6	16.9	11.4
16	16.0	15.8	18.7	12.8
18	17.6	17.0	20.4	14.2
20	19.2	18.2	22.1	15.6
22	20.8	19.4	23.8	17.0
24	22.4	20.6	25.5	18.4
26	24.0	21.8	27.3	19.8
28	25.5	23.1	29.0	21.2
30	27.1	24.3	30.7	22.7

Wt. Allowance: Wt. (lb.) Up to	Factor	Constant TMU
2.5	1.00	0
7.5	1.06	2.2
12.5	1.11	3.9
17.5	1.17	5.6
22.5	1.22	7.4
27.5	1.28	9.1
32.5	1.33	10.8
37.5	1.39	12.5
42.5	1.44	14.3
47.5	1.50	16.0

CASE AND DESCRIPTION
A Move object to other hand or against stop.
B Move object to approximate or indefinite location.
C Move object to exact location.

TABLE 8-3
Turn and Apply Pressure, T and AP

Weight	Time TMU for Degrees Turned										
	30°	45°	60°	75°	90°	105°	120°	135°	150°	165°	180°
Small— 0 to 2 Pounds	2.8	3.5	4.1	4.8	5.4	6.1	6.8	7.4	8.1	8.7	9.4
Medium—2.1 to 10 Pounds	4.4	5.5	6.5	7.5	8.5	9.6	10.6	11.6	12.7	13.7	14.8
Large— 10.1 to 35 Pounds	8.4	10.5	12.3	14.4	16.2	18.3	20.4	22.2	24.3	26.1	28.2
APPLY PRESSURE CASE 1—16.2 TMU. APPLY PRESSURE CASE 2—10.6 TMU											

TABLE 8-4
Grasp, G

Case	Time TMU	DESCRIPTION
1A	2.0	**Pick Up Grasp**—Small, medium or large object by itself, easily grasped.
1B	3.5	Very small object or object lying close against a flat surface.
1C1	7.3	Interference with grasp on bottom and one side of nearly cylindrical object. Diameter larger than ½″.
1C2	8.7	Interference with grasp on bottom and one side of nearly cylindrical object. Diameter ¼″ to ½″.
1C3	10.8	Interference with grasp on bottom and one side of nearly cylindrical object. Diameter less than ¼″.
2	5.6	**Regrasp.**
3	5.6	**Transfer Grasp.**
4A	7.3	Object jumbled with other objects so search and select occur. Larger than 1″ x 1″ x 1″.
4B	9.1	Object jumbled with other objects so search and select occur. ¼″ x ¼″ x ⅛″ to 1″ x 1″ x 1″.
4C	12.9	Object jumbled with other objects so search and select occur. Smaller than ¼″ x ¼″ x ⅛″.
5	0	Contact, sliding or hook grasp.

TABLE 8-5
Position, *P

CLASS OF FIT		Symmetry	Easy To Handle	Difficult To Handle
1—Loose	No pressure required	S	5.6	11.2
		SS	9.1	14.7
		NS	10.4	16.0
2—Close	Light pressure required	S	16.2	21.8
		SS	19.7	25.3
		NS	21.0	26.6
3—Exact	Heavy pressure required.	S	43.0	48.6
		SS	46.5	52.1
		NS	47.8	53.4

*Distance moved to engage—1″ or less.

TABLE 8-6
Release, RL

Case	Time TMU	DESCRIPTION
1	2.0	Normal release performed by opening fingers as independent motion.
2	0	Contact Release.

TABLE 8-7
Disengage, D

CLASS OF FIT	Easy to Handle	Difficult to Handle
1—**Loose**—Very slight effort, blends with subsequent move.	4.0	5.7
2—**Close**—Normal effort, slight recoil.	7.5	11.8
3—**Tight**—Considerable effort, hand recoils markedly.	22.9	34.7

TABLE 8-8
Eye Travel Time and Eye Focus, LT and EF

Eye Travel Time $= 15.2 \times \frac{T}{D}$ TMU, with a maximum value of 20 TMU.

where T = the distance between points from and to which the eye travels.
D = the perpendicular distance from the eye to the line of travel T.

Eye Focus Time = 7.3 TMU.

TABLE 8-9
Body, Leg, and Foot Motions

DESCRIPTION	SYMBOL	DISTANCE	TIME TMU
Foot Motion—Hinged at Ankle.	FM	Up to 4″	8.5
With heavy pressure.	FMP		19.1
Leg or Foreleg Motion.	LM —	Up to 6″	7.1
		Each add'l. inch	1.2
Sidestep—Case 1—Complete when leading leg contacts floor.	SS-C1	Less than 12″	Use REACH or MOVE Time
		12″	17.0
		Each add'l. inch	.6
Case 2—Lagging leg must contact floor before next motion can be made.	SS-C2	12″	34.1
		Each add'l. inch	1.1
Bend, Stoop, or Kneel on One Knee.	B,S,KOK		29.0
Arise.	AB,AS,AKOK		31.9
Kneel on Floor—Both Knees.	KBK		69.4
Arise.	AKBK		76.7
Sit.	SIT		34.7
Stand from Sitting Position.	STD		43.4
Turn Body 45 to 90 degrees— Case 1—Complete when leading leg contacts floor.	TBC1		18.6
Case 2—Lagging leg must contact floor before next motion can be made.	TBC2		37.2
Walk.	W-FT.	Per Foot	5.3
Walk.	W-P	Per Pace	15.0

TABLE 8–10
Simultaneous Motions

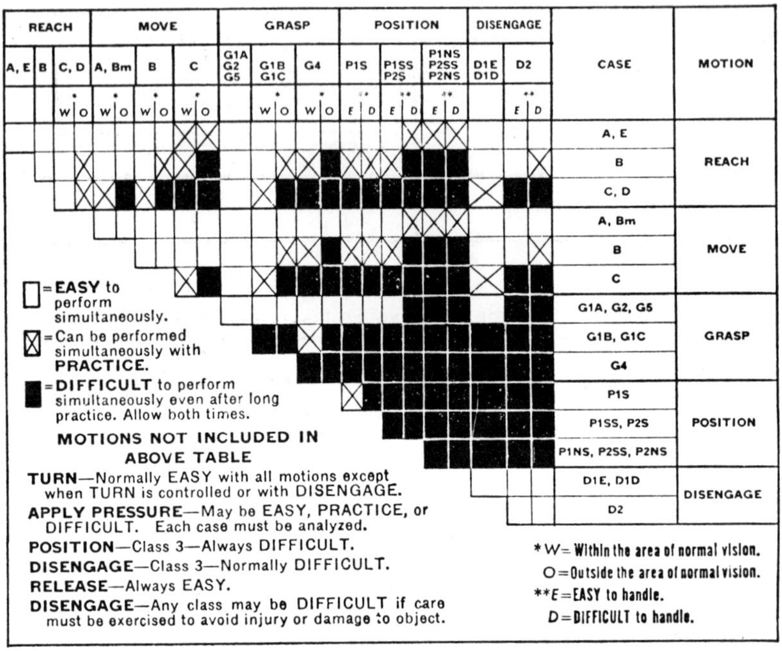

Precautions in Applying MTM

This chapter is intended to indicate the thoroughness and accuracy found in the successful application of predetermined motion time data. It is certainly not the proposal of this writing to prepare trained MTM without partaking of an MTM course coupled with concentrated applications experience.

It is not the intention to propose the supplanting of time study with MTM. The two complement each other very well in textile applications. Time study is foremost a measurement tool, whereas MTM is foremost a methods analysis tool. The establishment of a time standard through MTM is a mere by-product of its main application.

It is true, however, that until the advent of predetermined motion time standards the methods engineer had few engineering tools at his disposal. His science or art for the main part was dependent upon his past experience, developed through trial and error. The proper application of MTM in the textile industry will aid in the development of better standards through better methods.

Basic MTM Elements

The following is a general outline of the basic elements used in MTM to analyze motions of the hands, arms, and other parts of the body.

1. *Reach:* The basic element employed when the predominant purpose is to move the *hand* to a destination or general location. *Variables:*
   a. Length.
   b. Case (destination).
   c. Type (refers to the hand being in motion at the beginning and end of the reach).
2. *Move:* The basic element employed when the predominant purpose is to transport an object to a destination. *Variables:*
   a. Length.
   b. Case (destination).
   c. Type (refers to the hand being in motion at the beginning or end of the move).
   d. Weight factor.
3. *Turn:* The motion employed to turn the hand, either empty or filled, by a movement which rotates the hand, wrist, and forearm about the long axis of the forearm. *Variables:*
   a. Degrees turned.
   b. Weight factor.
4. *Apply Pressure:* The action employed to exert the additional force necessary to overcome the effects of resistance too great to be overcome by the normal move, turn, or crank. Apply

pressure is not a motion but rather is characterized by a hesitation or lack of motion.

5. *Grasp:* This is the basic element used when the predominant purpose is to secure sufficient control of one or more objects with the fingers to permit the performance of the next required basic element. There are fundamentally two different types of grasp:
   a. Pick-up grasp—the object is actually picked up. Sufficient control is obtained so that the object may be carried from one point to another.
   b. Contact grasp—the object is already limited in some way other than by the fingers. Because of this, the fingers need only contact the object to enable the hand to move it from one point to another.
6. *Position:* The basic element used to align, orient, and engage one object with another. The motions used are so minor that classification as other basic elements is not justified. *Variables:*
   a. Class of fit.
   b. Symmetry.
   c. Ease of handling.
7. *Release:* The basic element used to relinquish control of an object. Only two classes of release have been identified:
   a. Normal release—performed by opening the fingers as an independent motion. Employed usually to release an object obtained by a pick-up grasp and characterized by opening the finger.
   b. Contact release—when control of an object has been secured with a contact grasp, control is usually relinquished by employing a contact release. The release begins and is completed at the instant the following reach begins. Therefore, no measurable amount of time is consumed.
8. *Disengage:* This is the basic element used to break contact between one object and another. It includes an involuntary movement resulting from the sudden ending of resistance. *Variables:*
   a. Class of fit.
   b. Ease of handling.
   c. Care of handling.

9. *Eye Time:* Although the eyes are in constant use during nearly every work cycle, the time for moving and focusing them is usually a factor that is not limiting. In some cases this is not true, and we must be prepared with eye time data and a knowledge of how to use it when such cases occur. There are two types of eye time:
   a. Eye focus time—the time required to focus the eyes on an object and look at it long enough to determine certain readily distinguishable characteristics within the area which may be seen without shifting the eyes.
   b. Eye travel time—the time required to shift the eyes from one point to another. It is generally used in conjunction with eye focus time.
10. *Body, Leg, and Foot Motions.* Foot, leg, and body motions differ from arm and hand motions in that they are usually not used to perform motions which are accurate, probably because they are incapable of being closely controlled. Thus, they are not subject to as many variables affecting the performance time.
11. *Foot Motions.* Motions where the foot is moved with the ankle serving as a hinge or the instep serving as a fulcrum.
12. *Leg or Foreleg Motion.* Leg motions are movements of the foreleg or the entire leg where either the knee or the hip serves as a pivot.
13. *Sidestep.* The sidestep occurs when the body must be displaced sideways from one location to another in the immediate area without turning or taking more than one step. The length of a sidestep is measured from the center of the line.
14. *Turn Body.* This is a variation of the sidestep and occurs when turning the body to a new location while stepping from a work station.
15. *Bend.* This action occurs when the body is bowed from the waist. The upper portion of the torso is lowered to bring the hands within reach of an object that cannot be obtained when the body is held erect. Bending occurs only when it is made as a separate and distinct motion, with the destination of the hand at, or just below, the level of the kneecap.

16. *Stoop*. Begins with the downward movement of the body and ends when the hands touch the floor or an object close to the floor.
17. *Kneel on One Knee*. Begins with the downward movement of the body and ends as the knee touches the floor.
18. *Kneel on Both Knees*. Begins with a kneel on one knee, after which the remaining leg is lowered to its proper position.
19. *Sit*. Begins when the body is positioned at the location from which the body can be lowered to a chair, bench, or other support, without further movement of the body or feet. It ends when the body is seated.

The motions which are employed to return the body to the previous position after performing a bend, stoop, kneel on one knee, kneel on both knees, or sit are: arise from bend, arise from stoop, arise from kneel on one knee, arise from kneel on both knees, and stand.

20. *Walk*. This is applicable to short steps forward and backward, steps up and down stairs, carrying a load or without a load. In walking up or down stairs, if the risers are more than 12 inches high, the leg motion time should be used.

Although the measurement procedure used to establish standards for the body and leg motions was not as refined and accurate as that used in the development of the hand and arm standards, the data have proven satisfactory and will be used until more accurate studies have been made to sustain or modify them.

The Reach Element

As we have already learned, the time for a reach depends upon the length, the case (destination), and the type (continuity of motion).

Determining the Length of Reach. The greatest chance for error in applying MTM data to the reach motion lies in the determination of the length of the motion. In measuring, two precautions must be observed: (1) the true path of the motion must be measured, not just the straight line between the terminal points; and (2) the measurement must be made at the same point on the hand at the beginning and end of the motion—the knuckle at the base of the index finger is generally used.

Sometimes it becomes necessary to estimate distances that cannot be measured. In such a case, the observer should take great care. The maximum reach without body motion is about 30 inches. Since this is true, the MTM reach data table only shows values for reaches up to 30 inches. In case of a reach of more than 30 inches, it is necessary to extrapolate to find the value.

In the performance of *reaches* and *moves* it will be noted that in many cases the body bends or pivots to assist in the performance of these motions. In measuring the true distance of reach or move this assistance has to be removed.

In a "body bending assistance" first determine the distance moved by the hand and subtract from that the distance moved by the shoulder to find the chargeable distance moved by the hand. The point to measure the distance of assistance is about six inches from the center of the spine, just below the base of the neck.

In a "body pivot assistance," to determine the chargeable distance, measure the total distance moved by the hand. Then, subtract four times the distance moved from the point of measurement (six inches from the center of the spine and just below the base of the neck). If the arm is fullv extended, use a multiplication factor of 5 instead of 4.

Another classification is "wrist assistance." This action occurs when the hand pivots at the wrist in the direction of the *reach* or *move*. To find the chargeable distance of the reach or move, one would need to remove the distance traveled by the hand in pivoting.

Determining the Case of Reach. The time required to perform a reach is influenced by the nature of the object towards which the reach is made. The following classifications have been made:

Case A. Reach to object in a fixed location, or for object in other hand or on which other hand rests.

This is the fastest case of reach because the amount of conscious direction required to successfully perform the motion is reduced to a minimum. A common example of an object in a fixed location is the gear shifting lever on an automobile. The operator reaches repeatedly toward that control lever and, after a number of repetitions, he learns exactly where the lever is in relation to the rest of his body. When the reach is made toward an object held in the other hand, the same sense of orientation exists as when the reach is made toward an object in a fixed location. It can be classified as a Case A reach only if the hand resting on the object being reached is within three inches of the point to grasp by the other hand.

Case B. Reach for a single object in a location which may vary slightly from cycle to cycle.

In this instance, a certain amount of control is required to perform the motion successfully. Reaching for a wrench on one's workbench or a pencil in one's pocket are examples.

Case C. Reach for objects jumbled with other objects in a group so that a search and selection occur.

Case C reaches occur whenever material or other objects are jumbled together haphazardly, requiring selection before one of the objects can be grasped.

Case D. Reach for a very small object or where an accurate grasp is required.

When a reach is made for a very small object, the time for reach is increased due to the need for control over the motion. A reach for a common pin is an example of this. When a reach is made for an object that requires an accurate grasp, the time of the action is also slowed down. This will occur when reaching for fragile objects, or parts which require care to avoid injury.

Case E. Reach to indefinite location to get the hand in position for body balance or the next motion, or out of the way.

This case of reach occurs rarely in textile operations, but it is frequent on punch press operations where the hand is withdrawn from the die before any other motion is made.

Determining the Type of Reach (Continuity of Reach). The three types of reach are: (1) the hand is not in motion at the beginning nor at the end of the reach; (2) the hand is in motion at either the beginning or end of the reach; and (3) the hand is in motion at both the beginning and end of the reach.

The first type is known as the standard motion, since it occurs most frequently. The hand starts, accelerates, moves for a distance at a constant speed, decelerates, and stops. For the same length of motion, the second type is faster than the first because the time to start and accelerate, or the time to decelerate and stop are eliminated. The third type of motion is quicker than the others because of the elimination of the start and accelerate and the decelerate and stop.

Change direction is the basic motion employed to change the line or plane along which a reach is made. In order for a change direction to occur, the change must be abrupt enough to necessitate the application of an appreciable amount of muscular control. When a change direction occurs in a Case A reach, use the time for the next higher case, Case B, but do not show the reach as being a Case B reach. If there is a change direction encountered in any other case, use the time normally used for that particular case.

When recording reaches in a motion study one should first record the capital letter "R" to designate that the motion is a reach, next record the distance, and last the case. *Example:* reach 12 inches to a wrench lying on a workbench.

R12B (standard type 1 motion)

If the Reach is a type 2 motion, it would be written in the following manner:

mR12B (type 2 motion)

The small "m" preceding the "R" is indicative of the hand being in motion at the beginning of the reach. A type 3 motion would also have a small "m" following the case letter:

mR12Bm (type 3 motion)

Textile Mill Uses of Reach

Examples of reach, grasp, and move are most common in all operations throughout the textile mill.

Case A. Reach for an object in a fixed location or for an object in other hand or on which the other hand rests. *Examples:* (1) The action of the spinning doffer reaching to remove a full bobbin. Note: Through practice the doffer has learned to orient his body with each spindle so that the reach is performed without conscious directions. (2) The spooler tender reaching to remove a full cheese.

Case B. Reach for a single object in a location which may vary from cycle to cycle. *Examples:* (1) The weaver reaching for yarn ends to repair a warp break (ends can be easily grasped). (2) The spinner reaching for a full roving bobbin. (3) The drawing tender reach ing for a can of card sliver. (4) The battery tender reaching to secure an end of yarn on a bobbin (if the end is readily discernible and easily grasped).

Case C. Reach for objects jumbled with other objects in a group so that search and selection occur. *Example:* A battery tender reaching into an open type of battery buggy to secure bobbins. Note: If quills are prepositioned and gravity fed into the battery cart so that they may be easily grasped, then this reach becomes an example of Case A.

Case D. Reach for a very small object or where an accurate grasp is required. *Example:* A spooler tender reaching for an end of yarn on a warp bobbin (where the end is lying flat against the bobbin).

Case E. Reach to an indefinite location to get the hand in position for body balance, the next motion, or out of the way. *Example:* Positioning the hand preparatory to slamming off a loom, or preparing to use the hand or a hammer in disengaging a full cheese.

The Move Element

The time for a move depends upon four variables, namely, the length of the move, the case (destination), the type (continuity of motion), and the weight factor.

Determining the Length of a Move. In order to measure the distance moved, the procedure as outlined under reach should be observed.

Determining the Case of a Move. The time required to perform a move is influenced by the nature of the necessary control exerted to move an object from one point to another.

Case A. Move object to the other hand or against a stop.

When an object is moved to the other hand, a minimum of conscious direction is needed. This is classified as a Case A move because the sense of orientation ("kinesthetic sense") speeds up the motion.

When an object is moved against a stop, the need for positioning is reduced or eliminated, and the stop serves to decelerate the motion with a minimum of muscular control. An example of a Case A move of this type would be that of sliding a book against the back of a desk or closing a desk drawer.

Case B. Move an object to an approximate location.

A Case B move occurs when an object is moved to a general location, but where no special care has to be taken to line up the object exactly. An example of this occurs when an object is put aside in a container. Since stacking in an orderly manner or orientation is not required, the location is approximate and, therefore, a Case B move is employed.

Frequently objects are disposed of by tossing them aside. If the hand continues on in the same general direction after the object has been released without noticeable hesitation, a type 2, Case B move occurs. If the object is tossed aside and the hand stops and then starts off in another direction, the move should be classed as a standard (type 1) Case B move. If the latter occurs, care must be taken in measuring this move since the distance traversed by the hands does not necessarily terminate as the object is released. The entire distance through which the hand travels should be charged to the move.

Case C. Move object to an exact location.

When an object is moved to a location where it must be brought into exact relationship with another object, a Case C move occurs. A Case C move is often followed by a "position."

A common example of a Case C move is to move a piece of thread to the eye of an needle.

Determining the Type of Move. The required type of move is determined, as in the case of "reach," by analysis of the end result or the end position. Thus, a Case A move is limited to a move of an object from one hand to the other or against a fixed stop. A Case B move, which finds frequent industrial application, consists of the controlled movement of an object to an indefinite location. Movements categorized as Class C are directed to a precise location (within one-quarter inch), and always precede a "position."

As was discussed under "reach" there are three types of move: (1) the hand is not in motion at the beginning or at the end of the element; (2) the hand is in motion either at the beginning or at the end of the move; and (3) the hand is in motion both at the beginning and at the end of the move. The first type is the most common to all three cases of move.

Whether or not the hand is in motion depends upon the preceding and succeeding elements. All three cases (A, B, and C) may have the hand in motion at the beginning of the move, although in Case A and Case C moves the second type of move is rare. The motion of the hand at the end of a Case A move is sometimes practical where a hook grasp and sliding move against a stop have preceded the hand in motion. The Case C move, by definition, never allows the hand to be in motion at the conclusion of the move. A Case B move, however, lends itself to a pickup "on the fly" and a "toss aside" at the end of the move, and therefore, qualifies as a move of both types 2 and 3.

Determining the Weight Factor. When a lightweight object (2.5 pounds or less) is moved the weight does not affect the move time. If, however, the object weighs over 2.5 pounds the speed of the move will be reduced. The time values given on the data card must be adjusted to compensate for the weight involved. This is done by multiplying the basic move times by the appropriate multiplying factor, and adding to this the constant value. The multiplying factor and constant are found in the move table. When handling a weight with two hands, divide the weight by two and multiply by using the factor and the constant for half the original weight. The two basic components of the move with weight are (1) static, which is the constant and (2) dynamic, which is the factor times the table value.

The *static component* is the time required for the muscles of the hand and arm to take up the weight of the object. This occurs after the fingers have closed upon the object and it must take place before movement can begin.

The *dynamic component* is the time factor which, when multiplied by the TMU time, yields the extra time required to move the object.

Once an object has been picked up, do not allow the constant for any subsequent moves after the initial move involved in picking up the object. Also, do not allow the constant value for moves that are initiated when the hand is in motion (type 2 motion). The difference between male and female performance is chiefly significant in the movement of heavier weights.

In recording moves in a motion study, first set down the capital letter "M" to designate that the motion is a move, next record the distance the hand moves the object, and after this, record the case. *Example:* Move a wrench lying on a workbench 15 inches to another point on the workbench.

M15B

If the wrench weighed 5 pounds,

M15B5

If the wrench is dropped without a hand-halting motion,

M15B5m

If the wrench should happen to be prepositioned in such a way as to allow the wrench to be grasped and moved without the hand ceasing its forward motion, and the wrench can be dropped without the hand ceasing its forward motion, then,

mM15B5m

Textile Mill Examples of Move

Case A. Move object to other hand or against a stop. *Examples:* (1) The weaver throwing the shipper handle of a loom. (2) The spinning doffer moving an empty bobbin down the spindle to the stop. Note: It probably required a Case B move to position the bobbin over the spindle. (3) The spooling tender seating a starter on a spindle. To the beginner this "analysis" will appear "tight" due to the tolerances of fit. In practice, however, the starter is turned in an oblique path to the spindle to allow initial contact with the leading edge of the spindle case.

Case B. Move an object to an approximate location. *Examples:* (1) The spooling tender positioning a yarn end in the thread holder. (2) The batter tender laying an end across the battery hub. (3) The doffer placing an empty bobbin on a spindle.

Case C. Move object to an exact location. *Examples:* (1) A spinner making a piece-up. (2) The action of threading the eye of an needle. The C move in this case precedes a position. (3) A cloth inspector burling a delicate fabric.

The weight factor involved in the move element must be analyzed fully. It will occur when moving a card sliver can from the stand to the floor or when positioning a cloth roll onto a dolly. The tendency more often than not, is to over-correct for movements of weight.

Cranking

A cranking motion occurs when the hand follows a circular path and the forearm pivots about the elbow, acting as a ball and socket joint. The variables affecting a cranking motion are the diameter of the crank, and the resistance encountered in turning the crank.

The time required for a cranking motion is equal to:

$$R[(N \times T) + 5.2]$$

where R is the resistance factor (obtained from the move weight table), N is the number of turns made, T is the time allowed per turn (from Table 8-11), and 5.2 represents acceleration and deceleration.

TABLE 8-11
Time for Cranking Motions, Slight Resistance

Diameter of crank, inches	Leveled time, TMU: First revolution	Leveled time, TMU: Each additional revolution, T
1	13.7	8.5
2	14.9	9.7
3	15.8	10.6
4	16.6	11.4
5	17.3	12.1
6	17.9	12.7
7	18.4	13.2
8	18.8	13.6
9	19.2	14.0
10	19.6	14.4
12	20.2	15.0
14	20.7	15.5
16	21.2	16.0
18	21.6	16.4
20	21.9	16.7

Once the diameter of a crank exceeds 20 inches, the action involved is no longer a cranking motion, but rather, a move. In recording a cranking motion, first note the number of turns, next record "C" for crank, and last the diameter of the crank.

Example:

8C2

which indicates that there are eight turns of a 2 inch diameter crank. If more than 2.5 pounds or resistance are involved, then,

8C2 − 5

indicating that there are eight turns of a 2 inch diameter with 5 pounds of resistance encountered in the cranking motion.

Examples of cranking may be found in the lowering or raising of the immersion roller on a slasher, lowering cloth roll for doffing by using a crank, the winding of the traverse motion on a spinning frame, or the winding up of the rack during the doffing of a loom. Depending upon the body position of the battery tender, the winding of yarn about the battery hub could constitute a cranking motion.

The Turn Element

A turn is merely a special way of performing a reach or move. A separate table of values has been set up for turn because the motion is distinctive and requires a different amount of time to perform than the normal reach or move. The time to perform a turn depends on two variables, namely, the degrees turned and the weight factor.

Determining the Degrees Turned. To determine the degrees turned, observe one of the prominent knuckles of the hand as it rotates around the long axis of the forearm. It is usually practical to measure the degrees turned in steps of 15 degrees.

Determining the Weight Factor. The weight factor for a turn might be the weight of an object turned in free space or the resistance of an object (in pounds) supported (as a door handle). The three classifications of weight factors are: small–0 to 2 pounds; medium–2.1 to 10 pounds; and large–10.1 to 35 pounds.

Most reaches and moves have a turn internal to their completion. Seldom will a turn be found as a limiting motion. One example of a limiting motion might be rotating a warp bobbin about its long axis while seeking the end of the yarn.

In recording the turn element the degree of turn as well as the weight being handled must be noted as follows:

T15S-Turn hand 15°; 0 to 2 lb.
T15M-Turn hand 15°; weight of 2 to 10 lb.
T15L-Turn hand 15°; weight of 10 to 35 lb.

The Apply Pressure Element

Apply pressure occurs on a large number and variety of operations and can be easily overlooked by the inexperienced observer. However, it is easy to recognize since it is characterized by a distinct hesitation or lack of motion. It is found in actions such as the final tightening of a screw or in pushing or pulling.

Case AP1. Involves a minor readjustment of the body members followed by the actual application of pressure, such as removing a stuck bobbin in doffing.

Case AP2. Done without any preliminary readjustment of the body member, such as depressing a start button.

The Grasp Element

Grasp is the basic element used to secure control of an object so that the next motion can be performed. The variables affecting grasp are the type of grasp, and the conditions under which the grasp is made.

G1A. Small, medium, or large object by itself, easily grasped. Examples would be a small object on a table, a pencil lying on a desk, or a pen in a desk pen holder. When a weaver grasps the shipper lever on a loom he utilizes an G1A grasp even though the object is not picked up.

G1B. Very small object lying close against a flat surface. If the object is 1/8″ × 1/8″ or less in cross section G1B grasp is required. Picking up a needle or some object lying close to a flat surface, such as a yard stick, G1B grasp.

G1C. Interference with grasp on the bottom and on one side of a nearly cylindrical object, as illustrated in Figure 8-1. The three classes of this case of grasp

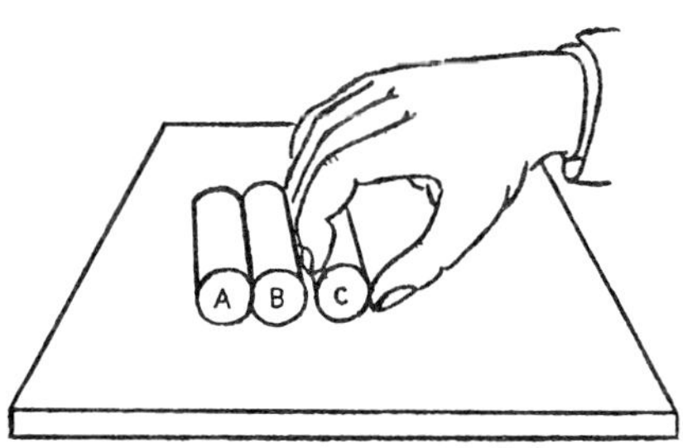

Figure 8-1. Example of the grasp element.

are: (1) the diameter is greater than $^1/_2$ inch, (2) the diameter is from $^1/_4$ to $^1/_2$ inch, and (3) the diameter is less than $^1/_4$ inch.

G2. Regrasp—a regrasp is accomplished by opening the fingers and (moving quickly so as not to lose control of the object) closing them again at, or near, the point of the original grasp. The purpose of regrasp is usually either to secure a better control of the object or to shift the object about in the hand.

G3. Transfer grasp—a transfer grasp takes place when an object is passed from one hand to the other with normal effort. The transfer grasp begins when the hand receiving the object grasps it and ends when the other hand releases it.

G4. The object is jumbled with other objects in a group so that search and selection occur. When objects are jumbled together so that they are interlocked and must be separated, the G4 grasp time does not apply since a multiplicity of motions must occur to separate them. The classes of grasp for this case are (1) objects larger than 1 × 1 × inch, (2) objects $^1/_4 \times ^1/_4 \times ^1/_8$ inch to 1 × 1 × 1 inch, and (3) objects smaller than $^1/_4 \times ^1/_4 \times ^1/_8$ inch.

G5. Contact, sliding, or hook grasp—the principal characteristic of this type of grasp is that the object is partially controlled and complete control is obtained merely by making contact with the object. A contact grasp is customarily employed when an object is to be slid along a surface.

Textile Examples of Grasp

The nature of the forms of textile packages permits most of them to be handled easily. Therefore, the predominant grasp is the G1A (e.g., grasping a bobbin, roving package, can of sliver, cheese, beam roll of cloth). Certain operations, however, require more complicated or extra motions than the mere pick-up grasp.

Examples: G1B (pinch grasp)—grasping yarn lying flat against a bobbin (spooler tender, battery tender); grasping the edge of cloth lying flat against a soft back stop. G1C—grasping stacked bobbins (stacked filling). G2 (regrasp)—battery tender gaining better control of bobbins in her hand; roll picker adjusting grip on cleaner. G3 (transfer grasp)—seldom used; always attempt to change method. G4 (when objects are jumbled to encourage search and selection)—battery tender grasping one, two, or more bobbins from a jumbled pile. G5 (contact sliding or hook grasp)—this grasp is frequently used on rough bobbins to obtain the yarn end (such as is done by the spooler tender, spinner, and battery tender). The initial contact with the bobbin is a contact grasp. The subsequent sliding motion utilizes the hand as a tool and is, therefore, a "move," the final motion is a contact release.

The Position Element

Position (P) is the basic element used to align, orient, and engage one object with another and where the motions used are so minor they do not justify classification as other basic elements.

Alignment: Brings the axes of both parts which coincide with the direction of travel into line with one another.

Orientation: Turns the entering part about the axis which coincides with the direction of travel.

Engagement: Enters one part into the other (distance moved to engage 1 inch or less).

Position can further be said to occur each time two objects are brought together where an accurate and predetermined relationship between the two objects is obtained. Two factors must be analyzed to classify position correctly:

1. The axis of travel—the common line along which the parts are engaged.

2. The point of initial engagement—the point along the axis of travel at which the object being positioned intersects the plane across the opening into which it is being positioned.

The three variables which affect position are class of fit, symmetry, and ease of handling.

Determining the Class of Fit. The class of fit affects the time required for aligning and engaging one object with another. The three categories that have been established are: *Class 1*—loose, or no pressure required; *Class 2*—close, or light pressure required; and *Class 3*—exact, or heavy pressure required. The rule of thumb limits for positioning at normal repetitiveness are: *Class 1*—$^1/_4$ to $^1/_{16}$ inch clearance; *Class 2*—$^1/_{16}$ inch to almost zero clearance; and *Class 3*—zero clearance.

These limits apply only to parts that are engaged. When positioning blind, usually allow the next higher class of fit than otherwise permitted.

Determining Symmetry. Symmetry is judged at the point of initial engagement. The three classes of symmetry are defined as follows:

Symmetrical (S): Object can be positioned in an infinite number of ways about the axis which coincides with the direction of travel (e.g., ○).

Semisymmetrical (SS): Object can be positioned in several ways about the axis which coincide with the direction of travel (e.g., □).

Nonsymmetrical (NS): Object can be positioned in only one way about the axis which coincides with the direction of travel (e.g., D).

In a multiple positioning operation, the second positioning would usually be symmetrical due to the constraining effect of the original positioning. If an object is guided or controlled mechanically, it may assume characteristics of semisymmetrical or symmetrical objects, insofar as positioning is concerned, although it may be nonsymmetrical in shape. In recording cases of position, show "P" with the degree of fit and the class of symmetry. For example, P1S indicates a loose position with symmetrical positioning and an object which is easy to handle.

Determining Ease of Handling. Ease of handling depends on how far the object is held from the point of initial engagement. Flexible materials are invariably difficult to handle.

Alignment occurs when an object is positioned to a point

or line. It is distinguished from the normal position as there is no engagement. The following are rules of thumb concerning align if an object is positioned within $^1/_2$ to $^1/_4$ inch (use "C" move); $^1/_4$ to $^1/_{16}$ inch (use P1SE or P1SD); and $^1/_{16}$ inch or less (use P2SE or P2SD).

If the object is small and is positioned to a line, use a single SS or NS positioning for the alignment. If the object is more than 2 inches long and it is aligned within $^1/_{16}$ inch, use two P2SE's or two P2SD's if two points have to be positioned.

Textile Examples of Position

Most "positions" which occur in the textile industry lie in the maintenance category. The following are examples of elements which must be carefully analyzed by the engineer to determine the type of position applied: (1) inserting reed hook through heddle (generally PINSD); and (2) threading the needle (generally PISE or PISD, depending upon how difficult it is to handle the thread).

The Release Element

Release is the basic element employed to relinquish control of an object. Only two types of release have been identified.

Case A. A normal release is performed by opening fingers as an independent motion. When an object has been secured by a pick-up grasp of any type, a normal release is usually employed to let it go.

Case B. A contact release is performed when control of an object is relinquished where originally control was gained with a contact grasp. A contact release consumes no measureable time. Most textile operations employ the normal release. The contact release is used in rare instances where a contact grasp precedes it—e.g., when smoothing cloth, the hand performs a contact grasp, a move (the hand is a tool), and a contact release.

The Disengage Element

The basic element employed to break the contact between one object and another is characterized by an involuntary movement resulting from the sudden ending of resistance. Three variables affect the time for disengagement, namely, the class of fit, the ease of handling, and the care in handling.

Class of Fit. *Loose* (D): Very slight effort when disengage blends with subsequent move.

Close (D2): Normal effort resulting in a slight recoil of not more than 4 to 5 inches. At the end of this recoil, the hand comes to a complete stop and may even return slightly in the direction opposite to the motion of the recoil.

Tight (D3): Considerable effort resulting in a marked recoil of 5 to 10, or even 12, inches.

The classes of fit are based on the assumption that no binding occurs during the disengage. If binding should occur, if D2 allow additional *Case B* of the "grasp" element, and if D3 allow additional *Case A* of the "apply pressure" element.

Ease of Handling. *Easy to Handle:* When objects can be grasped and disengaged without changing the grasp in any way.

Difficult to handle: When objects cannot be grasped and disengaged without changing the grasp.

Care in Handling. When care is exercised to avoid injury to the object or person: if normally D1 use D2; if normally D2 use D3; if normally D3 the motion pattern should be changed and this situation not allowed.

While the disengage action is encountered rarely in textile operations, it must be analyzed carefully when it does occur, such as in the sudden release of stuck bobbins in doffing or the breaking of yarn after the final wrapping around the battery hub.

The Eye Time Element

The two eye times are eye focus time and eye travel time.

Eye Focus Time. This is the time required to focus the eyes on an object and look at it long enough to determine readily distinguishable characteristics, within the area which may be seen without shifting the eyes. EF represents eye focus time.

Eye Travel Time. The time required to shift the eyes from one point or object to another, usually in conjunction with eye focus time. If the distance between two eye focuses is 10 inches and the perpendicular distance from the eye to the line of travel is 15 inches, then the eye travel time (ET) would equal T/D or 10/15 (see Figure 8-2).

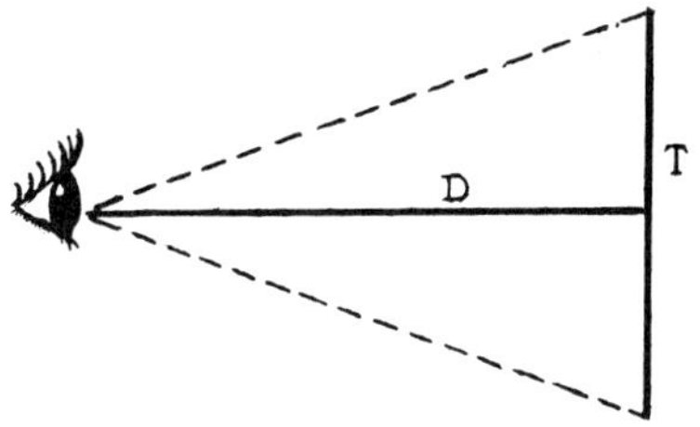

Figure 8-2. Eye focus and travel time.

The time allowed for eye travel is found by:

$$\text{Eye travel time allowed} = 15.2 \times T/D$$

where T is the distance traveled between eye focus points, and D is the perpendicular distance from the eye to the line of travel.

Eye travel can also be measured in degrees turned, allowing 0.285 TMU's per degree. The maximum eye travel time that should be allowed is 20.0 TMU's. Usually eye travel and eye focus are in connection with inspection jobs and where close observation is necessary, as in reading meters, watches, micrometers, or tachometers, or when performing an exact positioning.

Eye focus and travel are found to be limiting in certain inspection operations. Care must be exercised by the analyst to avoid excessive use of this generally internal motion.

The Walking Element

Fifteen TMU's per pace applies to short steps forward and backward, steps up or downstairs, and short starting and stopping steps.

The standard pace (without a load) appears to be 34 inches. For unobstructed walking while carrying a load allow 15.0 TMU's per pace for 0 to 5 pounds; 15.0 TMU's per pace for 5 to 35 pounds; 15.0 TMU's per pace for 35 to 50 pounds; and 17.0 TMU's per pace for 50 or more pounds.

The practitioner of MTM will find the discussion of walking somewhat confusing unless he recognizes the limitations of the classes of walk.

Unobstructive walking (15 TMU's per pace) is defined as walking on a hard, smooth, level surface while carrying a load that does not exceed 50 lb. An example is walking across a floor where the machinery is spaced at intervals that do not constitute a safety hazard.

Obstructive walking, on the other hand, requires that caution be exercised (e.g., walking on a slippery floor; between closely spaced moving machinery, etc.). Walking up and down aisles falls into this category, as well as walking across a floor which is cluttered with debris. A value of 17 TMU's per pace was found to be sufficient to cover this type of walk.

When pushing a two or four wheel truck, or buggy allow 17 TMU's per pace. In walking upstairs, if the risers are higher than 12 inches, use leg motion time. In recording walking, show the sequence given Table 8-12.

In the above analysis the walk is recorded under the right hand. Not only walk is recorded under the right hand but also all body motions (e.g., leg motion, L.M., and foot motion F.M.

TABLE 8-12[a]

Action	Left hand	Right hand	TMU's
Reach for object on table	R6B	R5B	8.6
Grasp object	G1A	G1A	2.0
Move object to body	M14B10	M14B10	20.1
Turn to start walking		TBC1	18.6
Walk to adjacent table		W5P	75.0
Move object to table	M14A10	M14A10	19.9
Release	RL1	RL1	2.0
Turn to start walking		TBC1	18.6
Walk between machines		W20PO	340.0

[a] For a key to all the letter symbols see Table 8-1 to 8-10.

The Principle of Limiting Motion

When two motions are performed at the same time, the total time required is that of the longer motion.

Combined motion—occurs when two motions are performed by the same body member at the same time.

R6B	8.6	TMU
~~T150S~~	~~8.1~~	TMU

(A diagonal line indicates that one element is internal to another, and is thus "lined out.")

This is a reach and arm turn. The R6B requires 8.6 TMU, and is, therefore, used in place of the T150S involving 8.1 TMU. The 150 degree turn was executed internal to the 6 inch reach, and was, therefore, limited.

Simultaneous motions—occur when two motions are performed by different body members at the same time (Table 8-13).

TABLE 8-13[a]

Left hand	TMU's	Right hand
	Example A	
R12B	14.2	R12C
G1A	3.5	G1B
	Example B	
M6C	10.3	M5B
P2SE	16.2	
RL1	15.2	ET15/15
	3.4	M1C
	16.2	P2SE
	2.0	RL1

[a] Control: Determine whether or not motions can or cannot be performed simultaneously.

Relatively uncontrolled motions: R-E, R-A, and M-A.

Partially controlled motions: R-B, M-B, and M-Bm.

Highly controlled motions: R-C, R-D, and M-C.

Table 8-10 shows simultaneous motions and motions that cannot be performed simultaneously.

The area of normal vision can be described as a 4 inch diameter circle located 16 inches from the eyes. The area of normal vision increases 1 inch with each 4 inch graduation from the eyes. Therefore, if the observer's eyes were 20 inches away from the work area, the area of normal vision would be 5 inches in diameter.

In practically all operations, walking plays a predominant part. The engineer is generally safe in using the per pace allowed time. This statement is true because the time per pace regardless of length differential is regarded as the same.

Care must be exercised in distinguishing between obstructive and unobstructive walking. Unobstructive walking is generally found on the spare floor where there is little danger of machine hazard, vibration, or moving equipment.

Obstructive walking is used when pushing a cart or walking between two close machines such as spinning frames or looms.

PART II

SYSTEMS, PROCEDURES, AND CONTROLS

CHAPTER 9

Job Evaluation

Introduction

The purpose of job evaluation is to set up pay ranges for various jobs in a plant by allowing credit for the mental and physical efforts required. Evaluation of the individual should, of course, be avoided.

By comparing the requirements of one job to another, on a scientific basis, the relative value of each may be determined accurately and be classified by means of a point system. Every job is made up of three major factors, namely, job prerequisites, job deterrents, and job responsibilities, and it is readily apparent that different jobs involve varying percentages of each of these factors. It is up to the industrial engineer to determine, from the type of work performed in the plant, the proper percentage of the total to be allocated to each of the three major factors. This can be done by evaluating 10 or 15 jobs that are in line with the wages paid in the industry in general. A good cross section of low, medium, and high paying jobs should be included. Trial weightings to determine the distribution of the factors should be made, until weightings resulting in the lowest mean deviation are found. A distribution for textile mill jobs of 40 per cent for prerequisites, 30 per cent for deterrents, and 30 per cent for responsibilities may be used as a starting point for setting up job values (as shown in Table 9-1).

Having determined the relative values of the jobs with the use of a job evaluation plan, it is possible to correct any inequities revealed at some future time when general wage increases are made, thereby gradually installing proper wages for all jobs. It is the industrial engineer's duty to keep the system up-to-date and

TABLE 9-1
Factors and Percentage-Point Values

Factors	Percentage
Training and skill	
1. Trade knowledge	10
2. Experience	10
3. Skill	10
4. Initiative and ingenuity	10
	—
Total	40
Physical and mental requirements	
5. Mental effort	5
6. Physical effort	10
	—
Total	15
Responsibility	
7. Material or product	10
8. Equipment or process	10
9. Safety of others	5
10. Work of others	5
	—
Total	30
Job conditions	
11. Working conditions	10
12. Hazards	5
	—
Total	15

to make suitable revisions when job contents are changed or new processes are added.

For purpose of illustration we will use a total of 500 points, distributed among each of the subfactors as shown in Table 9-2, with each subfactor divided into five degrees or steps. It may be found desirable to subdivide certain of these factors into 10 degrees, if local conditions warrant it.

When setting up the definitions for each factor, it is necessary to make them as clear, accurate, and concise as possible in order to avoid the possibility of crediting point values several times for the same job requirement. The outline of each degree should also be clear and definite, and since no single set of definitions could cover all of the requirements of every job, it is necessary in many instances to use comparisons between known job requirements. As an instance, a sweeper would be rated low in

TABLE 9-2
Factors, Degrees, and Point Values

Factors	Degree of factor				
	5th	4th	3rd	2nd	1st
Training and skill, 40%					
1. Trade knowledge, 10%	50	40	30	20	10
2. Experience, 10%	50	40	30	20	10
3. Skill, 10%	50	40	30	20	10
4. Initiative and ingenuity, 10%	50	40	30	20	10
Physical and mental requirements, 15%					
5. Mental effort, 5%	25	20	15	10	5
6. Physical effort, 10%	50	40	30	20	10
Responsibility, 30%					
7. Material or product, 10%	50	40	30	20	10
8. Equipment or process, 10%	50	40	30	20	10
9. Safety of others, 5%	25	20	15	10	5
10. Work of others, 5%	25	20	15	10	5
Job conditions, 15%					
11. Working conditions, 10%	50	40	30	20	10
12. Hazards, 5%	25	20	15	10	5

most factors, whereas a loom repairer would probably rate high. Other jobs could be expected to fall between these two extremes.

Job Description and Rating

Printed forms should be prepared on which the job may be clearly and accurately described, including what is to be done, the skill required, and the method to be used. The regular duties should be described together with the material used or produced, the equipment or machinery operated or serviced, the tools used, the mental or physical requirements, the variable working conditions, the safety equipment supplied, and the safety hazards involved in performing the job. The more details listed in the job description, the easier will it be to correctly rate the job.

1. Trade Knowledge

This factor measures the mental development or technical knowledge requirements of the job that are not developed by actual work on the job, but are acquired from schooling or vocational training.

1st Degree: Low degree of technical knowledge. Read and write, add and subtract whole numbers.

2nd Degree: Below average degree of technical knowledge. Use simple measuring instruments and read simple drawings. Add and subtract decimals and fractions.

3rd Degree: Average degree of technical knowledge. Use fairly complicated drawings, formulas, and a variety of precision measuring instruments.

4th Degree: Above average degree of technical knowledge. Use complicated drawings, advanced mathematics, and a wide variety of precision measuring instruments.

5th Degree: High degree of technical knowledge. Must be able to handle the most complicated mechanical, electrical, or other engineering problems.

2. *Experience*

This factor measures the time required to learn the work properly. This should include the time on jobs of a similar nature and job training where this procedure is followed, but should not include the trade knowledge or mental development required, which was measured and credited under *Trade Knowledge.*

1st Degree: Low degree of experience (up to 3 months).
2nd Degree: Below average degree of experience (3 months to 1 year).
3rd Degree: Average degree of experience (1 to 3 years).
4th Degree: Above average degree of experience (3 to 5 years).
5th Degree: High degree of experience (over 5 years).

Note: The first or low degree should be adjusted to the time allowed a trainee in the plant, with each succeeding degree prorated upward.

3. *Skill*

This factor measures the amount of physical skill, precision, and control required on the job, including the required coordination of several senses (e.g., setting equipment with gauges when feel, vision, and hearing are used).

1st Degree: Low degree of skill. Use most simple tools. Labor-type work.

2nd Degree: Below average degree of skill. Use simple tools with moderate skill on rough tasks.

3rd Degree: Average degree of skill. Use most common tools on normal work. Some precision and control required.

4th Degree: Above average degree of skill. Use most tools with skill on regular work. Use regular gauges for some precision work where close control is required.

5th Degree: High degree of skill. Use tools on work requiring close expert attention. Use various gauges requiring close tolerances and advanced skill.

4. *Initiative and Ingenuity*

This factor measures the amount of independent action and use of judgment required by the job. If decisions are to be made or if directions are provided for all phases of the work, the required degree of this factor will vary. A simple job may require little initiative and ingenuity, whereas jobs of a more complex nature may require considerable work analysis.

1st Degree: Little initiative and ingenuity. Requires only ability to follow simple orders and make a few simple decisions.

2nd Degree: Below average degree of initiative and ingenuity. Follow more detailed instructions and make some minor decisions by using a moderate amount of judgment.

3rd Degree: Average degree of initiative and ingenuity. Requires ability to operate where some ability and planning are involved. Ordinary decisions must be made to assure proper operation and production of good quality products.

4th Degree: Above average degree of initiative and ingenuity. Requires considerable ability and planning with many decisions to be made.

5th Degree: High degree of initiative and ingenuity. Requires outstanding ability and extensive planning to meet varying conditions on a complex job.

5. *Mental Effort*

This factor measures the mental or visual effort requirements of the job, in terms of varying degrees of concentration of thought necessary to operate the equipment or perform the job properly, and the amount of visual attention needed.

1st Degree: Low degree of mental effort. Little mental or visual attention required on job where work has a long or slow cycles.

2nd Degree: Below average degree of mental effort. Requires frequent mental and visual effort on job where the flow of work is intermittent, requiring attention only at intervals.

3rd Degree: Average degree of mental effort. Requires continuous mental and visual effort on work requiring constant attention.

4th Degree: Above average degree of mental effort. Requires concentrated mental and visual attention for extended periods.

5th Degree: High degree of mental effort. Requires highly concentrated and exacting mental and visual attention, with considerable foresight and planning.

6. *Physical Effort*

This factor measures the amount of physical fatigue resulting from lifting, pushing, or pulling various weights of materials in normal or strained positions, for varying periods of time.

1st Degree: Low degree of physical effort. Little or no physical effort where operator sits, stands, or moves in a normal manner.

2nd Degree: Below average degree of physical effort. Work is with lightweight material, or at times with materials of average weight. Work is done mostly in a normal position with occasional lifting or moving of lightweight materials in a strained position.

3rd Degree: Average degree of physical effort. Requires continuous work with lightweight materials, often in a strained position. May involve loading and unloading a series of machines in a continuous cycle.

4th Degree: Above average degree of physical effort. Requires considerable physical effort in handling average weight materials in a strained position, or frequently handling heavy materials in a normal position.

5th Degree: High degree of physical effort. Requires continuous physical effort in handling heavyweight materials, often in a strained position.

7. Responsibility for Material or Product

This factor measures the amount of responsibility for preventing loss, through damage, of the material produced. Consideration must be given to the salvage value of any damage, in addition to the possibility of total loss.

1st Degree: Low degree of responsibility for material or product. Possibility of damage is slight, requiring little attention. If damage occurs, it is minor.

2nd Degree: Below average degree of responsibility for material or product. Damage to material or product is possible but is easily avoided with some attention. If damage occurs, it is small.

3rd Degree: Average degree of responsibility for material or product. Damage to material or product is possible, requiring close attention some of the time. Moderate loss from damage is possible.

4th Degree: Above average degree of responsibility for material or product. Close attention is required most of the time to prevent damage to material or product. Loss from damage and the cost of salvage are high.

5th Degree: High degree of responsibility for material or product. Damage to material or product is likely unless extremely close attention is given. Potential losses are very high.

8. Responsibility for Equipment or Process

This factor measures the responsibility for preventing loss through damage to the equipment or for loss in production due to breakdown of equipment resulting from lack of attention.

1st Degree: Low degree of responsibility for equipment or process. Possibility of damage is slight requiring little attention. If damage occurs, it is easily repaired and loss is negligible.

2nd Degree: Below average degree of responsibility for equipment or process. Damage to equipment or process is possible, but is easy to avoid with some attention. If damage occurs, it is minor.

3rd Degree: Average degree of responsibility for equipment or process. Damage to equipment or process is possible, requiring close attention some of the time. Some loss from damage is possible.

4th Degree: Above average degree of responsibility for equipment or process. Close attention is required most of the time to prevent damage to equipment or process. Loss from damage and the cost of salvage is high.

5th Degree: High degree of responsibility for material or product. Damage to equipment or process is likely unless extremely close attention is given. The loss, if it occurs is very high.

9. Safety of Others

This factor measures the possibility of injury to other workers by the worker on the job. Consideration should be given to the extent of the injury which might result from carelessness or negligence.

1st Degree: Low degree of responsibility for safety of others. Little chance of causing injury to others either because the work is done in an isolated place or no moving machine is involved.

2nd Degree: Below average degree of responsibility for safety of others. With reasonable care, accidents causing injury to others can be avoided. If they occur, injuries involve minor cuts or bruises.

3rd Degree: Average degree of responsibility for safety of others. Must follow all safety rules to prevent lost-time accidents to other workers.

4th Degree: Above average degree of responsibility for safety of others. Must be constantly alert to prevent serious injury to other workers due to hazards connected with the job. Other workers may avoid injury by being alert.

5th Degree: High degree of responsibility for safety of others. Carelessness may result in fatal accidents, so that worker must use every precaution.

10. Work of Others

This factor measures the job requirement for helping, directing, or guiding the work of other workers. The degree of responsibility is based on the number of other workers involved.

1st Degree: Low degree of responsibility for work of others. The worker works alone and is responsible only for his own job.

2nd Degree: Below average degree of responsibility for work of others. Works in group with one or two helpers, assisting and guiding them.

3rd Degree: Average degree of responsibility for work of others. Is responsible for helping or setting up the work for a group of up to 10 other workers.

4th Degree: Above average degree of responsibility for work of others. Is responsible for helping or setting up the work for a group of up to 25 other workers.

5th Degree: High degree of responsibility for work of others. Is responsible for helping or setting up the work for a group of over 25 workers.

11. Working Conditions

This factor measures the conditions in the area where the work is to be performed, and the extent to which these conditions may make the work disagreeable. The amount of lint, dirt, dust, heat, cold, dampness, oil, grime, and the time of exposure should be considered.

1st Degree: Low degree of disagreeable work conditions. Good working conditions without any disagreeable elements

2nd Degree: Below average degree of disagreeable work conditions. Fair working conditions with some dust, but condition is not continuous.

3rd Degree: Average degree of disagreeable work conditions. Some disagreeable working conditions due to lint, dust, dirt, heat, or cold, but only one or two of these conditions are present continuously.

4th Degree: Above average degree of disagreeable work conditions. Dirt, dust, grime, heat, or cold may prevail most of the time.

5th Degree: High degree of disagreeable work conditions. Continuous exposure to several disagreeable conditions.

12. Hazards

This factor measures the accident and health hazards connected with the job, with the degree or seriousness of injury also being considered.

1st Degree: Low degree of hazards. Little chance of injury from accidents—no health hazards.

2nd Degree: Below average degree of hazards. Accidents unlikely and health hazards few. Injuries, if they do occur, involve minor cuts or bruises.

3rd Degree: Average degree of hazards. Must follow safety rules and use precautions to avoid accidents that result in lost time.

4th Degree: Above average degree of hazards. Great care must be used to avoid serious injuries.

5th Degree: High degree of hazards. Extreme care must be taken to avoid accidents that could result in loss of limbs or eyes.

Figure 9-1 illustrates a form that may be used for writing the job description and summarizing or totaling the point values.

Figure 9-1. Job Evaluation Rating Form

Job title:	Job no.:	Dept.:
No. of employees:	Male:	Female:
Present hourly rate:	Evaluated rate:	Job grade:

Job Description and Tools Required

Summary of Job Point Value

Factor	Degree	Points	Basis of assigned degree
Trade knowledge			
Experience			
Skill			
Initiative and ingenuity			
Mental effort			
Physical effort			
Material or product			
Equip. or process			
Safety of others			
Work of others			
Work conditions			
Hazards			
Total			

It may be desirable to prepare an additional form to be used for outlining, in more detail, the substantiating data for determining

TABLE 9-3
Present Jobs, Hourly Rates, and Point Values

Department	Operator	Number of 40 hour operators	Rate per hour, dollars	Evaluated points
Opening and picking	Opener tender	3	1.13	270
	Bale opener	4	1.08	240
	Picker tender	3	1.13	280
	Utility man	3	1.15	300
Carding	Card tender	15	1.17	245
	Card grinder	5	1.36	370
	Lap man	3	1.08	175
	Oiler and cleaner	3	1.08	215
	Drawing tender	9	1.17	270
	Slubber tender	15	1.28	270
	Section man	3	1.36	310
Spinning	Spinner	60	1.15	245
	Doffer	30	1.17	275
	Roll picker	6	1.08	190
	Roving boy	9	1.08	175
	Oiler	3	1.08	210
	Tape boy	3	1.10	220
	Sweeper and cleaner	9	1.00	155
	Section man	9	1.36	315
Spooling and warping	Spooler tender	24	1.13	230
	Warper tender	6	1.15	220
	Warper creeler	3	1.10	210
	Yarn boy	3	1.08	210
	Section man	3	1.40	385
Weaving	Slasher tender	3	1.36	315
	Slasher helper	3	1.30	250
	Tie-in machine operator	3	1.34	275
	Beam man	3	1.08	240
	Warp man	3	1.08	260
	Weaver	24	1.24	285
	Sweeper and cleaner	3	1.00	155
	Loom fixer	18	1.40	390
Cloth room	Burler and inspector	6	1.17	240
	Folder machine operator	3	1.20	270
	Shearing machine operator	3	1.20	275
	Utility man	3	1.30	245
	Sweeper and cleaner	3	1.00	255
	Section man	3	1.36	290
	Total	315		
	Average		1.18	256

TABLE 9-4
Average Hourly Rates for Various Point Values

Job title[a]	Point value	No. of workers	Rate per hour	Total cost per hour	No. of workers, same points	Total cost	Average rate per hour
Loom fixer	390	18	$1.40	$25.20	18	$25.20	$1.40
Section man (BC)	385	3	1.40	4.20	3	4.20	1.40
Card grinder	370	5	1.36	6.80	5	6.80	1.36
Section main (spin)	315	9	1.36	12.24	12	16.32	1.36
Slasher tender	315	3	1.36	4.08			
Section man (card)	310	3	1.36	4.08	3	4.08	1.36
Utility man (O & P)	300	3	1.15	3.45	3	3.45	1.15
Section man (cloth)	290	3	1.36	4.08	3	4.08	1.36
Weaver	285	24	1.24	29.76	24	29.76	1.24
Picker tender	280	3	1.13	3.39	3	3.39	1.13
Doffers (spin)	275	30	1.17	35.10			
Tie-in machine operator	275	3	1.34	4.02	36	42.72	1.19
Shearing machine operator	275	3	1.20	3.60			
Opener tender	270	3	1.13	3.39			
Drawing tender	270	9	1.17	10.53	30	36.72	1.22
Slubber tender	270	15	1.28	19.20			
Folding machine operator	270	3	1.20	3.60			

Warp man (weave)	260	3	1.08	3.24	3	3.24	1.08
Slasher helper	250	3	1.30	3.90	3	3.90	1.30
Card tender	245	15	1.17	17.55			
Spinner	245	60	1.15	69.00	78	90.45	1.16
Utility man (cloth)	245	3	1.30	3.90			
Bale opener	240	4	1.08	4.32			
Beam man (weave)	240	3	1.08	3.24	13	14.58	1.12
Burler and inspector	240	6	1.17	7.02			
Spool tender (BC)	230	24	1.13	27.12	24	27.12	1.13
Tape boy (spin)	220	3	1.10	3.30			
Warp tender (BC)	220	6	1.15	6.90	9	10.20	1.13
Oil and clean (card)	215	3	1.08	3.24	3	3.24	1.08
Oiler (spin)	210	3	1.08	3.24			
Warp creeler (BC)	210	3	1.10	3.30	9	9.78	1.09
Yarn boy (BC)	210	3	1.08	3.24			
Roll picker (spin)	190	6	1.08	6.48	6	6.48	1.08
Lap man (card)	175	3	1.08	3.24	12	12.96	1.08
Roving boy (spin)	175	9	1.08	9.72			
Sweep and clean (spin)	155	9	1.00	9.00			
Sweep and clean (weave)	155	3	1.00	3.00	15	15.00	1.00
Sweep and clean (cloth)	155	3	1.00	3.00			

[a] BC refers to the Barber-Colman spooler, and O & P refers to the opening and picking operations.

degree values for each factor. On the illustrated rating form, under the column "Basis of Assigned Degrees," values ranging from low to high may be noted, whether or not an additional substantiating data sheet is used. Close cooperation between the production department supervisors and the industrial engineer is required throughout the evaluation process and all information collected in respect to job contents should have the approval and agreement of all concerned. In this way, many labor inequities may be uncovered and corrected, as the evaluation progresses.

Converting Point Values to Money Values

The typical list of job rates and number of employees in Table 9-3 may be used to illustrate the method followed in this outline to convert the point values to money values. Table 9-3 also shows present jobs, number of workers, rate per hour, and the evaluated point value for each. It is now necessary to es-

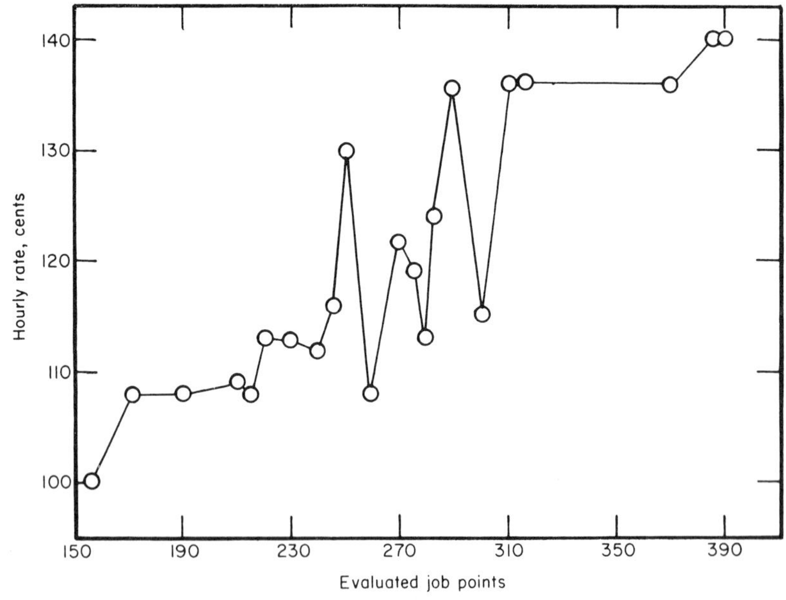

Figure 9-2. Evaluated job points and present pay rates.

tablish weighted average hourly rates for the various point values as outlined in Table 9-4.

Several formulas may be used to readjust the hourly rates to their proper relationship to point values. Reference to Figure 9-2, which is a graph of the average wage rates for various point values, indicates that several highly rated jobs are paid inadequately and that several of the lower rated jobs are overpaid. These inequities may be corrected with the use of a calculated line of average relationship through the plotted points on the diagram, letting job point values represent the X axis and job rates the Y axis. Tabulate this data as given in Table 9-5 and

TABLE 9-5
Calculation of Wage Line, Representing Line of Average Relationship of Job Points to Average Pay Rate per Hour

	Evaluated job points, X	Hourly pay rate, Y, dollars[a]	XY	X^2
	390	1.40	546.00	152100
	385	1.40	539.00	148225
	370	1.36	503.20	136900
	315	1.36	428.40	99225
	310	1.36	421.60	96100
	300	1.15	345.00	90000
	290	1.36	394.40	84100
	285	1.24	353.40	81225
	280	1.13	316.40	78400
	275	1.19	327.25	75625
	270	1.22	329.40	72900
	260	1.08	280.80	67600
	250	1.30	325.00	62500
	245	1.16	284.20	60025
	240	1.12	268.80	57600
	230	1.13	259.90	52900
	220	1.13	248.60	48400
	215	1.08	232.20	46225
	210	1.09	228.90	44100
	190	1.08	205.20	36100
	175	1.08	189.00	30625
	155	1.00	155.00	24025
Totals	$\Sigma X = 5860$	$\Sigma Y = 26.42$	$\Sigma XY = 7181.65$	$\Sigma X^2 = 1644900$

[a] These rates are based on 1957 wage scales.

then calculate the equation of the wage line as shown by Illustrative Example 9-1.

ILLUSTRATIVE EXAMPLE 9-1
Calculation of the Equation for the Wage Line

$$C = [\Sigma XY - (\Sigma X)(\Sigma Y)/N]/[\Sigma X^2 - (\Sigma X)^2/N] \quad (1)$$
$$= [7181.65 - (5860 \times 26.42/22)]/[(1644900) - 5860^2/22]$$
$$= 0.00172, \text{ where } N \text{ equals the number of point values.}$$

$$b = (\Sigma Y - C\Sigma X)/N \quad (2)$$
$$= (26.42 - 0.00172 \times 5860)/22$$
$$= 0.7428$$

The line of average relationship, the wage line, or the regression line is found from:

$$Y = b + CN \quad (3)$$
$$= \$0.7428 + 0.00172N$$

Example: If N equals 300 points, then
$$Y = \$0.7428 + 0.00172 \times 300$$
$$= \$1.26$$

Plotting the Wage Chart

On suitable graph paper, as shown in Figure 9-2, label the X and Y axes appropriately. Next, determine the spread of points to be used, which in this case is from 140 to 390. The payroll listed 13 different hourly rates. After evaluation it was found that there were 22 different point values among the jobs. It would not be practical, usually, to suggest 22 rates, and it would be advantageous, especially to the payroll department, to have a minimum number of rates. With this thought in mind, determine the number of steps or labor grades (this value will generally range between 6 and 9 rates).

Because the highest labor grade is $1.40 per hour and, as the formula for the wage line indicates, 380 points equals $1.40, we can consider 380 points to be the top grade. Since 140 points represents the lowest grade, we need to determine only the number of intermediate grades. Adding units of 40 points will give us seven uniform grades or steps which, when plotted on the graph in Figure 9-3, indicate a much more equitable distribution of

Figure 9-3. Wage line, evaluated job points, and labor grades assigned (represented by the circled values).

wages than the wage chart in Figure 9-2. Table 9-6 lists the seven labor grades, the point range of each, and the respective wage rates.

TABLE 9-6
Labor Grades, Point Range, and Hourly Rates

Labor grade	Point range	Hourly rate, dollars
1	140 to 175	.98[a]
2	180 to 215	1.05
3	220 to 255	1.12
4	260 to 295	1.19
5	300 to 335	1.26
6	340 to 375	1.33
7	380 to 415	1.40

[a] One dollar is the minimum rate.

The Cost of Applying Evaluated Rates

Generally, the cost for changing to evaluated rates is small, since it merely redistributes the current wages in a more proper manner. In Table 9-7, the method for determining the change in cost if the evaluated rates were to be put into effect is shown.

TABLE 9-7
Comparison of Labor Costs

Department	Job title	Labor grade	Point value	Actual rate, dollars	Evaluated rate, dollars	Change cost, dollars +	Change cost, dollars −	No. of workers	Change cost, dollars per hour +	Change cost, dollars per hour −
Opening and picking	Opener tender	4	270	1.13	1.19	.06		3	.18	
	Bale opener	3	240	1.08	1.12	.04		4	.16	
	Picker tender	4	280	1.13	1.19	.06		3	.18	
	Utility man	5	300	1.15	1.26	.11		3	.33	
Carding	Card tender	3	245	1.17	1.12		.05	15		.75
	Card grinder	6	370	1.36	1.33		.03	5		.15
	Lap man	1	175	1.08	1.00		.08	3		.24
	Oiler and cleaner	2	215	1.08	1.05		.03	3		.09
	Drawing tender	4	270	1.17	1.19	.02		9	.18	
	Slubber tender	4	270	1.28	1.19		.09	15		1.35
	Section man	5	310	1.36	1.26		.10	3		.30
Spinning	Spinner	3	245	1.15	1.12		.03	60		1.80
	Doffer	4	275	1.17	1.19	.02		30	.60	
	Roll picker	2	190	1.08	1.05		.03	6		.18
	Roving boy	1	175	1.08	1.00		.08	9		.72
	Oiler	2	210	1.08	1.05		.03	3		.09
	Tape boy	3	220	1.10	1.12	.02		3	.06	
	Sweeper and cleaner	1	155	1.00	1.00			9		
	Section man	5	315	1.36	1.26		.10	9		.90

Spooling and warping	Spooler tender	3	230	1.13	1.12		.01	24		.24
	Warper tender	3	220	1.15	1.12		.03	6		.18
	Warper creeler	2	210	1.10	1.05		.05	3		.15
	Yarn boy	2	210	1.08	1.05		.03	3		.09
	Section man	7	385	1.40	1.40			3		
Weaving	Slasher tender	5	315	1.36	1.26		.10	3		.30
	Slasher helper	3	250	1.30	1.12		.18	3		.54
	Tie-in machine operator	4	275	1.34	1.19		.15	3		.45
	Beam man	3	240	1.08	1.12	.04		3	.12	
	Warp man	4	260	1.08	1.19	.11		3	.33	
	Weaver	4	285	1.24	1.19		.05	24		1.20
	Sweeper and cleaner	1	155	1.00	1.00			3		
	Loom fixer	7	390	1.40	1.40			18		
Cloth room	Burler and inspector	3	240	1.17	1.12		.05	6		.30
	Folder machine operator	4	270	1.20	1.19		.01	3		.03
	Shearing machine operator	4	275	1.20	1.19		.01	3		.03
	Utility man	3	245	1.30	1.12		.18	3		.54
	Sweeper and cleaner	1	155	1.00	1.00			3		
	Section man	4	290	1.36	1.19		.17	3		.51
Totals (based on 1 hour on each of 3 shifts)									2.14	11.13

Net change cost = $11.13 − 2.14 = $8.99
Present cost per payroll hour = $373.67
Evaluated cost per payroll hour = $364.68
Savings per payroll hour = $8.99 (2.4%)
Net savings per hour = $3.00

In this instance, the change cost would equal a saving of \$3.00 per hour or 2.4 per cent of the total payroll.

Use of Job Evaluation

Unlike piece rates or job assignments, which may be corrected immediately if they are found to be wrong, pay rates cannot be changed overnight, without causing considerable trouble. If every rate were to be increased, doubtlessly this could be done readily. However, to lower some rates and raise others requires time. The best time to make adjustments is when general wage adjustments are made, and it may require several general adjustments to correct some of those rates that were extremely out of line.

One of the most useful purposes of the job evaluation plan is its application when new jobs are started. Its use when new jobs are analyzed and rated is far superior to the old "rule of thumb" method previously used.

CHAPTER 10

A Textile Cost System

Introduction

The basic principle of a textile cost system is to control, by predetermined budget estimates, the manufacturing costs of various yarns and materials. A reconciliation of the manufacturing cost should be made periodically to determine basic trends in operating efficiency of the various departments, and to indicate when cost estimates need revision. Revisions are necessary when general wage increases are effected, and when major production changes occur. With predetermined departmental labor and overhead costs, it is possible to accurately estimate yarn or fabric costs for quotations when bidding for orders. This chapter outlines the steps necessary for setting up a cost system for a yarn or fabric mill.

Weekly Production Schedule

The first step in setting up a cost system is to determine the weekly production upon which the costs will be based. This may be pounds or yards of yarn and fabric, dozens of socks, sheets, or blankets, or any other units produced in a 40, 80, or 120 hour week.

Department Production and Waste Divisors

Working back from pounds of finished product, the production requirements of each process or department may be determined, and at the same time waste divisors may be calculated for use in determining the costs to be allowed for subsequent waste. For illustrative purpose, the productive departments and waste percentages shown in Table 10-1 (based on 100,000 pounds of weekly fabric production) may be used.

TABLE 10-1
Production and Waste Data—Weaving Mill

Item	Production, lb.	Waste		Waste divisor[a]
		Per cent	Pounds	
Cotton opened	114,130			0.8762
Opening and picking waste		3.00	3325	
Picker production	110,800			0.9025
Card waste		6.00	6650	
Card production	104,150			0.9601
Drawing waste		0.50	550	
Drawing production	103,600			0.9652
Intermediate waste		0.50	550	
Intermediate production	103,050			0.9704
Spinning waste		0.75	830	
Spinning production	102,220			0.9783
Spooling waste		0.25	280	
Spooling production	101,940			0.9810
Warping waste		0.25	280	
Warping production	101,660			0.9837
Balance waste		1.50	1660	
Weaving production	100,000			1.0000

[a] Waste divisor = weaving production, lb./production at each process, lb.

List of Machinery

The number of productive machines needed to turn out the finished product must next be determined. From the production setup for each department, the number of machines may be calculated and a list of the machinery prepared, together with the motors required. Replacement values and horsepower requirements may be established from such data.

List of Departments

A list of departments may then be set up, consisting of each process or series of processes for which there is measurable production in total units (e.g., hanks, tridents, beams, picks, or yards in such processing departments as opening and picking, carding, drawing, fly frames, spinning, spooling, warping, winding, slashing, tieing-in, weaving, and the cloth room). It is also necessary to include departments for such expenses as power, heating, handling, storage, repairing, and roll covering, for the redistribution of these costs to the productive departments

Floor Plan

With a list of machinery and a floor plan of the mill, the square feet required for each process or department are determined by actual measurement. The total value of the buildings divided by the total square feet of area yields the value per square foot. From this, the values by processes or departments are established readily.

Normal Weekly Payroll

Setting up a weekly payroll budget can be accomplished by using average earnings for piece work and hourly rates for day-pay labor for each department. Based on the production schedules and the machinery assigned to each department, the number of workers are determined and a standard or normal weekly payroll is drawn up.

General Labor and Overhead Budget

The general labor and overhead budget should include allowances for such items as taxes, salaries, office expenses, insurance, depreciation of buildings and machinery, and administrative expenses. These items may be determined readily from the company books, but other categories of the overhead budget will require considerable study of plant records. Included are such variable items as roll covering, fuel, power, sizing materials, supplies, building and machinery repairs, and water. These expenses are determined on a weekly basis for distribution to the various departments as shown in Table 10-2.

TABLE 10-2

Expense	Basis
Steam	Floor area of each department
Power	Horsepower of each department
Repair shop	Repair work of each department
Supplies	Estimates used in each department
Depreciation of building	Value of departmental floor area
Depreciation of machinery	Value of machinery

Distribution of Labor and Overhead

On a large sheet of paper, with columns headed by department names, the weekly payroll is posted for both piece work and hourly labor, together with the general and overhead budget amounts per week. These departmental budgets are shown for piece work, day work, and overhead, and are used in determining departmental production costs for the various styles.

Departmental Production Costs

Forms may be prepared for each process or department, listing the grain sliver, hank roving, yarn numbers, fabric styles produced, and the weekly output of each. Piece work and day work labor costs per pound for each style are added and totaled with overhead cost, to arrive at the departmental production cost as outlined in Figures 10-1 to 10-9.

Figure 10-1. Departmental Production Costs—Opening through Drawing Operations

Required production, pounds per 80 hr.: Grain sliver delivered:

Drawing Frame Tender Piece Rates

Standard per frame	Rate per frame hank
Piece work labor per 80 hr.	
Day work labor per 80 hr.	
Overhead per 80 hr.	

Production Cost per Pound

Grain sliver	Piece work labor	Day work labor	Total labor	Overhead	Total

Formulas for Calculating Production Costs

Piece work labor cost per pound

$$= \frac{\text{piece rate per frame hank} \times 7000}{\text{delivery per frame} \times \text{grain sliver} \times 840 \times \text{waste divisor}^{a}}$$

$$\text{Day work labor cost per pound} = \frac{\text{day work labor cost per 80 hr.}}{\text{pounds per 80 hr.} \times \text{waste divisor}}$$

$$\text{Overhead cost per pound} = \frac{\text{overhead cost per 80 hr.}}{\text{pounds per 80 hr.} \times \text{waste divisor}}$$

[a] The waste divisor is used to allow for waste in subsequent processing.

Figure 10-2. Departmental Production Costs—Intermediate Operations

Spindles per frame: Frames in place: Spindles in place:

Hank roving produced	Pounds per spindle, per 80 hr.	Total pounds required	No. of 80 hr. spindles required
1.			
2.			
etc.			
Total			

Tender Piece Rates

Hank roving produced: Rate per frame hank:

Standard piece work labor per week: $

$$\text{Cost per spindle-week} = \frac{\text{standard day work labor per week, \$}}{\text{no. of spindles required}}$$

$$= \frac{\text{standard overhead per week, \$}}{\text{no. of spindles required}}$$

Production Cost per Pound, $

Hank roving	Piece work labor	Day work labor	Total labor	Overhead	Total

Formulas for Calculating Production Costs

$$\text{Piece work labor cost per pound} = \frac{\text{piece rate per frame hank} \times \text{hank roving}}{\text{spindles per frame} \times \text{waste divisor}^{a}}$$

$$\text{Day work labor cost per pound} = \frac{\text{day work labor cost per spindle-week}}{\text{pounds per spindle-week} \times \text{waste divisor}}$$

$$\text{Overhead cost per pound} = \frac{\text{overhead cost per spindle-week}}{\text{pounds per spindle-week} \times \text{waste divisor}}$$

[a] The waste divisor is used to allow for subsequent waste in processing.

Figure 10-3. Departmental Production Costs—Spinning Warp or Filling

Production Data

	Yarn no.	Total pounds	Pounds per spindle per 80 hr.
Required 80 hr. production			

Spindles per frame: Frames in place: Spindles in place:
Spindles required for production:

Piece Rates for Spinners and Doffers

Yarn no.	Rate per frame hank	
	Spinner	Doffer

Standard piece work labor per 80 hr. (spinners and doffers): $

$$\text{Cost per spindle per 80 hr.} = \frac{\text{standard day work labor per 80 hr., \$}}{\text{total no. of spindles required}}$$

$$= \frac{\text{standard overhead per 80 hr., \$}}{\text{total no. of spindles required}}$$

Production Cost per Pound, $

Yarn no.	Spinner labor	Doffer labor	Day work labor	Total labor	Overhead	Total

Formulas for Calculating Production Costs

$$\text{Spinners or doffers cost per pound} = \frac{\text{piece rate per frame hank} \times \text{yarn no.}}{\text{spindles per frame} \times \text{waste divisor}}$$

$$\text{Day work labor cost per pound} = \frac{\text{day work labor cost per 80 hr. spindle}}{\text{pounds per 80 hr. spindle} \times \text{waste divisor}}$$

$$\text{Overhead cost per pound} = \frac{\text{overhead cost per 80 hr. spindle}}{\text{pounds per 80 hr. spindle} \times \text{waste divisor}}$$

Figure 10-4. Departmental Production Costs—Barber-Colman Spooling

	Yarn no.	Total pounds	Pounds per spindle per 80 hr.
Required 80 hr. production			

Spindles per frame: Spindles per side:
No. of spindles required for production:

Piece Rates for Spooler Tenders

Yarn no.	Pounds per trident	Rate per trident

Standard piece work labor per 80 hr. (spooler tenders): $

$$\text{Cost per spindle per 80 hr.} = \frac{\text{standard day work labor per 80 hr., \$}}{\text{total no. of spindles required}}$$

$$= \frac{\text{standard overhead per 80 hr., \$}}{\text{total no. of spindles required}}$$

Production Cost per Pound

Yarn no.	Piece work labor	Day work labor	Total labor	Overhead	Total

Formulas for Calculating Production Costs

$$\text{Spooler tender cost per pound} = \frac{\text{rate per trident}}{\text{trident, lb.} \times \text{wasted divisor}}$$

$$\text{Day work labor cost per pound} = \frac{\text{day work labor cost per 80 hr. spindle}}{\text{pounds per 80 hr. spindle} \times \text{waste divisor}}$$

$$\text{Overhead cost per pound} = \frac{\text{overhead cost per 80 hr. spindle}}{\text{pounds per 80 hr. spindle} \times \text{waste divisor}}$$

Figure 10-5. Departmental Production Costs—Barber-Colman Warping

Required 80 hr. production:	No. of warpers:
No. of warper hours required:	Pounds per warper hour:

Piece Rates for Warper Tenders and Creelers

Yarn no.	Ends per beam	Yards per beam	Pounds per 1000 yards[a]	Rate per 1000 yards, $	
				Warper tender	Warper creeler

Standard piece work labor per 80 hr. (warper tenders and creelers): $

$$\text{Cost per warper hour} = \frac{\text{standard day work labor per 80 hr.,}}{\text{no. of warper hours required}}$$

$$= \frac{\text{standard overhead per 80 hr., \$}}{\text{no. of warper hours required}}$$

Production Cost per Pound

Yarn no.	Tender labor	Creeler labor	Day work labor	Total labor	Overhead	Total

Formulas for Calculating Production Costs

Warper tender or creeler cost per pound

$$= \frac{\text{rate per 1000 yards}}{\text{pounds per 1000 yards} \times \text{waste divisor}}$$

$$\text{Day work labor cost per pound} = \frac{\text{day labor cost per warper hour}}{\text{pounds per warper hour} \times \text{waste divisor}}$$

$$\text{Overhead cost per pound} = \frac{\text{overhead cost per warper hour}}{\text{pounds per warper hour} \times \text{waste divisor}}$$

[a] Pounds per 1000 yards = (ends per beam × 1000)/(yarn no. × 840)

Figure 10-6. Departmental Production Costs—Slashing

Required 80 hr. production, yards of slashed warp:
Slashers in place: Yards per slasher hour:

Standard day labor per 80 hr.:
Standard overhead per 80 hr. (without sizing material):

Production Cost per Yard of Cloth

Cost per yard of cloth = labor costs/yards of cloth per 80 hr.

= overhead costs/yards of cloth per 80 hr.

Production Cost per Pound of Cloth

Cost per pound of cloth = cost per yard of cloth × yards per pound of cloth

Sizing Material Cost

Sizing material cost ($ per pound of warp yarn) = sizing material cost/warp yarn produced, lb.

Sizing material cost ($ per pound of cloth) = sizing material cost per pound × % warp yarn in cloth

Figure 10-7. Departmental Production Cost—Tieing-In

Required 80 hour production of warp yarn:
Standard day labor per 80 hours, $:
Standard overhead per 80 hours, $:

Production Cost per Pound of Warp Yarn

$$\text{Cost per pound of warp yarn} = \frac{\text{cost of day labor}}{\text{pounds of warp yarn per 80 hours}}$$

$$= \frac{\text{overhead costs}}{\text{pounds of warp yarn per 80 hours}}$$

Total: labor + overhead = cost per pound of warp yarn

Cost per Pound of Cloth for Tieing-In

To find the cost per pound of cloth for tieing-in, multiply the total cost per pound of warp yarn for tieing-in by the percentage of warp yarn in the cloth.

Cost per pound of cloth = cost per pound of warp yarn tied-in × warp yarn in cloth, %

Figure 10-8. Department Production Cost—Weaving

Style:	Width:	Sley:	Picks:
Warp counts:	Number of ends:		
Yards per pound:	Filling counts:		

Per Cent Warp and Filling

$$\text{Warp yarn, pounds per yard, \%} = \frac{\text{number of ends} \times (1 + \text{contraction, \%})}{\text{count} \times 840}$$

$$\text{Filling yarn, pounds per yard, \%} = \frac{\text{picks per inch} \times \text{reed width}}{\text{count} \times 840}$$

Sizing material:

Totals: warp + filling + size = 100%

Loom: Picks per minute: Looms per weaver:

Production per 80 Hour Loom

Pounds per 80 hour loom

$$= \frac{\text{P.P.M.} \times 60 \times 80 \times \text{efficiency, \%}}{\text{P.P.I.} \times 36} = \text{yards/yards per pound}$$

$$\text{Pounds per 1000 picks} = \frac{\text{pounds per yard} \times 1000}{\text{P.P.I.} \times 36}$$

Production Cost per Pound

Cost per pound, weaver = (piece work, cost per 1000 picks/pounds per 1000 picks)
Cost per pound, battery hand = (piece work, cost per 1000 picks/pounds per 1000 picks)
Cost per pound, loom fixer day work = (cost per 80 hour loom/pounds per 80 hour loom)
Cost per pound, weaving day work = (cost per 80 hour loom/pounds per 80 hour loom)
Cost per pound, weaving overhead = (cost per 80 hour loom/pounds per 80 hour loom)

Figure 10-9. Department Production Cost—Cloth Room

Standard cloth production per 80 hours, yards:
Standard day work labor per 80 hours, $:
Standard overhead per 80 hours, $:

Production Cost per Yard

Cost per yard, labor = cost/yard per 80 hours
Cost per yard, overhead = cost/yard per 80 hours
Total = labor + overhead
Cost per pound = cost per yard × yards per pound of cloth

Summary Sheets

Summary sheets may be prepared for labor, overhead, and total costs per pound for the various rovings, as well as for warp and filling yarns for application in development of the cost cards. The factors comprising a cost card are shown in Figure 10-10.

Figure 10-10. Standard Cost Card Data

Standard Cost Card

Style: Width: Sley: Picks: Yards per pound:
Warp yarn: Number of ends: Looms used:
Filling yarn: Picks per minute:

Standard Cost per Pound[a]

Element	Labor	Over-head	Total
Warp yarn (cost per pound × warp, %)			
Filling yarn (cost per pound × filling, %)			
Slashing (cost per pound × yards per pound)			
Sizing (cost per pound × warp, %)			
Tying-in (cost per pound × warp, %)			
Weaving piece work (cost per 1000 picks/ pounds per 1000 picks)			
Battery hands piece work (cost per 1000 picks/ pounds per 1000 picks)			
Weaving, day work (cost per 80 hr. loom/ pounds per 80 hr. loom)			
Weaving, overhead (cost per 80 hr. loom/ pounds per 80 hr. loom)			
Cloth room (cost per yard × yards per pound)			
Total production cost per pound			
Cotton cost (cost per pound/waste divisor)			
Total mill cost per pound			

[a] To the above standard cost may be added the expected mark-up or profit plus the selling expenses to determine the cost for sales purposes.

In order to determine the trend of actual production cost versus standard cost, the comparison shown in Figure 10-11 may be made.

Figure 10-11. Cost Reconciliation

Style	Construction	Pounds produced	Total standard cost per pound	Estimated production cost
Totals				
Actual weekly labor cost:				
Standard weekly overhead:				
Total weekly cost:				

Concluding Notes

Some cost systems are set up with the departmental labor separated into three groups; direct labor piece work, direct labor day work, and indirect labor day work. Direct labor day work may be performed by workers who handle definite pounds or other units per day (e.g., roving hoisters and yarn boys). On the other hand, section men, sweepers, spindle oilers, and band boys, would be listed as indirect labor. This permits easier checking and control of the number of workers used or required, and may result in a more accurate division of departmental costs when a number of different counts of yarn are run.

Other systems apply a multiplier to the direct labor cost to determine the indirect and overhead charges to be used. The multiplier is determined by dividing the combined departmental total indirect labor and overhead costs (for a specified period) by the total direct labor costs for the same period. This

TABLE 10-3

Item	Roving type	
	10 × 5 inch	12 × 6 inch
Direct labor cost per 1000 lb.	$20.50	$18.50
Indirect labor and overhead at 1.50 times direct labor cost	30.75	27.75
Total spinning cost per 1000 lb.	$51.25	$46.25

method may result in great variations in estimated costs, as illustrated in Table 10-3 for a case history involving an indirect labor and overhead multiplier of 1.50 for spinning, with several of the same counts spun from both 10 × 5 inch roving at a direct labor cost of $20.50 per 1000 pounds, and from 12 × 6 inch roving at a cost of $18.50 per 1000 pounds. Table 10-3 is a comparison of the department costs for these two similar counts.

Obviously, the direct labor cost of the yarn from the 12 × 6 inch roving is less than that from the 10 × 5 inch roving, but the indirect and overhead costs have remained about the same because the spindle and frame production remains unchanged.

CHAPTER 11

Materials Handling

Introduction

Materials handling is a form of transportation which usually identifies the systems and modes of conveyance of goods from place to place within a factory. The development and planning for efficient in-plant movement and storage of materials usually is, or should be, the responsibility of the industrial or materials handling engineer in the textile industry. Although the discussion below refers to a finishing plant, similar applications in spinning and weaving mills will become apparent.

For industry in general, it is estimated that the average materials handling costs comprise approximately 22 per cent of total labor costs. This all-industry average is on the low side for textile finishing mills. For example, in one modern finishing mill the obvious materials handling cost amounts to about 30 per cent of the total labor cost, and hidden costs should bring this value up to approximately 35 to 40 per cent.

The ratio of materials handling cost to total labor costs will vary from mill to mill, and in a textile finishing mill this item represents one of the largest labor expenses of the entire operation. This expense does not transform products in any way, or make them more valuable to the customer, and is thus a nonrecoverable expense which is necessary to facilitate productive operations.

Preliminary Survey

Materials handling systems and equipment applications are similar to the effective use of standards, in that the benefits received in these areas are proportionate to the efficiency of the installation and the effectiveness of maintenance. Mills regularly spend money and time installing and maintaining machinery and equipment and consider this normal. They will usually adopt this same outlook with respect to the installation and maintenance of standards and materials handling.

To initiate an over-all materials handling survey, it is suggested that the survey be accomplished by a materials handling engineer. For major projects and extensive programs materials handling consultants are available. Large textile firms will probably find it economical to employ one or more materials handling engineers on a full time basis. Medium or small size firms must rely on industrial engineers and the materials handling surveys and information provided by equipment manufacturers and distributors. It is important that these surveys, provided in conjunction with sales of equipment, be carefully evaluated to insure that one is buying the equipment best suited for each individual problem.

When industrial engineers engage in materials handling activities, it is suggested that the mill's engineering department cooperate closely on the projects. The engineering department contributes efficient data and decisions concerning the physical characteristics of plants, machinery, and equipment, and is vital to a well engineered materials handling installation.

There is no magic formula to solve materials handling problems; in fact, the only common denominators in materials handling are the techniques of analysis and the equipment used. A system or equipment that will operate at low cost in one plant is not necessarily economical in another. However, one can materially profit by successful concepts and programs in other plants.

Selling the Project

In common with other tools of industrial engineering, the way to an effective materials handling project is through the effective selling of ideas. These ideas will often meet with severe resistance from the very people to be helped. The operator or worker resistance is not the major hurdle. The average mill worker will usually adopt a new method, if he learns that his work will be made easier as a result of methods change. This can and should be demonstrated. Management and supervision are the people to whom the industrial engineer must first sell a materials handling project. Therefore, before initiating a project it is well to consider how to present the findings of a materials handling survey. It is suggested that from the very beginning the industrial engineer work closely with management and supervision, with the objective of having these gentlemen identify themselves with the project and understand the over-all and individual problems and their solutions. The contributions to the project by supervision must be freely acknowledged. It is not to be inferred that the responsibility of recommendations is passed from the engineer to supervision or to management. The responsibility of the presentation and possible failure of the installation belongs to the engineer. The credit for successful installation belongs to the team who cooperated and worked together to achieve the cost reduction.

In addition to the normal resistance to change that is expected from any individual, materials handling projects often meet additional resistance because of the overlapping of responsibility for materials handling between supervision and the project engineer. This condition is normal and is to be anticipated and understood by the engineer. Working closely with supervision on the development of the project and acknowledging that the results of this work are owing to teamwork and cooperation, will help alleviate this overlapping of responsibility.

Materials handling and industrial engineering are staff functions and as such do not have line authority to directly implement recommendations. Staff functions guide and facilitate line operation and are not responsible for daily operation. In addition to the training in problem solving and measurement, this ability of the industrial engineer to concentrate on one individual problem and not be diverted by daily operation is one of the major advantages used by the industrial engineer in problem solving. The success of all such work depends upon the decisions and efforts of other people. Therefore, the selling of sound ideas is the key to successful materials handling. The planning of this selling approach should receive appropriate attention at the very beginning of project work and should continue as the project is developed and finished.

Analysis of Handling Problems

Cost reduction materials handling projects begin with an analysis of present methods, controls, and procedures. It is suggested that a quick and rough analysis of operation and process flow be initiated in order to determine the degree of potential savings. This preliminary survey provides the first contact with supervision on a project. Sitting down with a supervisor and permitting him to describe and explain the operation that is thoroughly familiar to him, starts both parties working together. Other sources for the same information are available, but they do not provide the same opportunity for establishing an atmosphere of cooperation and teamwork.

If this preliminary survey, which incidentally should not take over a day or two, does indicate potential savings, then a more thorough analysis of the present operation is required. First, the characteristics of the material must be determined, measured, and evaluated. The sizes and weights, the degree of protection required, susceptibility to dirt and dust, crushing resistance, and stability must be studied and reliable data obtained. After determining the characteristics of the goods to be handled, more elaborate and detailed process flow charts and diagrams should be developed. These charts and diagrams should cover both the physical operation and the required paper work. This determination of where and in what sequence the goods flow immediately must be supported by analysis covering quantity moved, distance traveled, method of travel, frequency of movement, who and what departments perform the moves, whether the movement is scheduled or initiated as required, the time required for movement (in both clock and man hours), historical and standard costs, physical characteristics of areas involved, whether area is high or low cost, and finally, machine speeds, feeds, and capacity. It is not suggested that every problem be given this depth of study, however, materials handling, in common with other industrial engineering facilities, requires judgment as to the amount of effort required to economically solve a problem.

When the amount of background information pertaining to the present method has been developed, the principles of motion economy should be applied to the system as a whole and then to the sections and components of the system. The questions asked in the motion economy approach are:

1. Can we eliminate? It is necessary that the operation or movement be performed, and is it productive?

2. Can we change the sequence and reduce required movement?

3. Can we combine operations or sections of operations in order to reduce movement?

4. Can the operation or movement be simplified?

In materials handling one of the first attempts at simplification is the increasing of unit volume loads. The more material moved at one time, the fewer required moves. The application of mechanical or electrical equipment to replace human effort is another primary objective.

An important component in the development of a new materials handling method is the experimenting and testing to insure that the new techniques are practical prior to the purchase of equipment. Arrangements can often be made with suppliers and distributors of materials handling equipment for obtaining sample facilities for these trials. It is strongly recommended that before purchasing substantial portions of materials handling equipment samples be obtained, and then a pilot shipment be purchased in order to insure that one is getting the type of equipment needed.

Next, one should project the expected costs with use of the new method. An accurate comparison to the present cost will develop expected savings. Justification and amortization of the required materials handling equipment, and planned reconditioning and rearranging of the plant area, will develop the period of time required to pay for the investment out of operating savings.

Handling Equipment

The types of materials handling found in a finishing mill may be roughly classified into two categories. These are the in-plant movement of materials from place to place, and processing movement or machine travel. An integral component of materials handling, especially in a finishing mill, is the permanent and temporary storage of goods.

Platform and box hand trucks, pallets, racks, conveyors, J-boxes or scrays, batch roll trucks, fork lift trucks and stackers, and manual and power hoists are some of the equipment available for individual plant materials handling. This equipment may be purchased in many sizes, capacities, and several power types. Most of the heavier and more expensive equipment is designed for general industry, but is readily adaptable to textile handling.

The importance of the platform and box hand trucks in today's material handling planning should not be overlooked. The ease of mobility of hand trucks is an important asset for short distance moves during in-process handling and temporary storage of cloth.

The flexibility, speed, and economies of electric, gasoline, and propane powered fork lift trucks have greatly benefitted the textile industry. Through the use of lift trucks equipped with self-stacking pallets or pallets and racks many fabric storage and transporting problems have been alleviated. The characteristics of the material to be moved and stored, the normal production quantities moved as a production unit, and the physical characteristics of the plant largely determine the type of fork truck and pallets most economical for a given operation. Generally speaking, substantial horizontal distances indicate the use of counterbalanced fork trucks owing to their speed. Where space is restricted and horizontal movement not too great, straddle type fork trucks have an advantage because they generally require less aisle space to operate.

Self-stacking pallets permit the more efficient utilization of space, but have a low degree of selectivity. Pallets and tiered racks offer 100 per cent selectivity, but are not as efficient in space utilization as self-stacking pallets.

The advantages, disadvantages, and prices of the various types of materials handling equipment must be balanced with the requirements of the individual plant operation. Equipment should be selected for over-all economy and versatility. Economies in initial cost can be more than offset by high maintenance expense.

Continuous processing in the finishing range eliminates or simplifies manual handling between machines. J-boxes or scrays, creepers, and other types of mechanical cloth storage assist in the balancing of individual operations. The processing or machine travel movement can pass cloth through machines, dryers, and ranges, in the direction of flow with a minimum of labor expense. However, machine travel of cloth and range alignment of machinery can increase labor expense if it is necessary to reprocess the cloth through the same operation. Also, if the reduction in operating speed of individual units in the range offsets the savings in manual handling, or if sufficient processing flexibility is lost then it is not economical to insert the processing equipment in range. These are problems of the materials handling or industrial engineer.

In addition to the handling of cloth, finishing mills require movement of substantial quantities of chemicals, dyestuffs, and other processing materials. If processing materials are used in liquid form, consideration should be given to dissolving or mixing at a central location and pumping through pipes to the various points of use.

Dry chemicals, coal, and wrapping materials can be handled economically by fork trucks or conveyors, and movement of these materials should be considered. Packaging is another major materials handling area that is rapidly becoming a specialized branch of materials handling, utilizing the services of a packaging engineer.

Summary

In summary, the substantial materials handling expense in a textile finishing mill requires this area to receive its proper share of managerial and industrial engineering attention. The selling of materials handling ideas and projects should start with the project itself and continue through to management and supervision and down to the mill worker who must perform the task. Thorough analysis of present methods and operation is required both for the development of improved methods and for justification of investment in new equipment and the expense of plant layout rearrangement. The attention, selling, teamwork, and analysis required for successful materials handling improvement is of little value without successful equipment selection.

CHAPTER 12

Automation in the Textile Industry

Introduction

For management and labor alike "automation" has sometimes appeared like a cloud on the horizon of industrial production. Employees seeking to maintain their status have seen the machine replacing men to the detriment of both the individual and the laboring community as a whole. Management, faced with increasing costs all along the line and keener competition for the consumers' dollar, looks on the situation with an eye to the capital needs involved. To remain competitive, the advantages of automation must be attained, but the goal requires tremendous capital outlays.

Definition

What is this new method created by the engineering minds to serve the mass production techniques of our industries? Automation, by its connotation is tied to the word automaton. An *automaton* is, according to Webster, "a self-moving machine —one which imitates the motions of men. . ." The world would be a very backward place today without machines which imitate the motions of man. To be precise, no one would want to see a world devoid of automatons, but if such a situation existed it would not be so for long. Man did not begin to progress from the stone age until he had discovered and learned to use three things: fire, the lever, and the wheel. Two of these were automatons in the sense that they were machines that replaced men. Thus, automation is but a continued refinement of man's development from the stone age, which has lifted mankind from the realm of purely animal life to the standards we enjoy today. Automation is merely the application of machines to the tasks of men, doing the work faster and cheaper, so that more wealth is created for all. With this in mind, let us take a look at automation in the textile industry.

Automation in the Textile Industry

Complete automation is virtually nonexistent in the textile industry as a whole. It is granted that the word automation is applied somewhat loosely at times to include all applications where mechanical devices are used to supplant human effort, and it is in this sense only that this chapter can deal with the advancements in the field of automation in the textile industry.

The day when an operator can sit at a console, and, by manipulating various levers and buttons, blend raw stock, card, draw, spin yarn, and weave finished fabric is a long way off. However, the steps necessary to establish this level of automation are being considered, and, to a limited extent, are being adopted.

Great strides have been made in the introduction of automation to certain specific areas. Most of these efforts have been in the direction of increased unit capacity, with little progress in combining or automating cycles. Mechanical aids to human effort do not constitute automation in its true sense. Only when cycles are integrated and combined does the machine begin to supplant the human factor.

Here then is the heart of the matter: If the industry is slow to automate its processes, what has been, or is being done to mechanize? The following discussion will consider applications of automation in various areas of textile production.

Materials Handling

One of the most fertile fields for mechanization is the handling of materials. The days of manually handling large packages, such as bales of cotton, are gone. Motorized or electric jeeps load and unload, stack, and distribute faster and more economically than can be done by hand. Throughout the plant, hoists, conveyors, and mechanized dollies transport materials between operations, more easily and more cheaply than can be done with manpower.

Let us take a look at some of this materials handling equipment, to see what it has accomplished in the mills themselves. The handling of raw materials comes first. In the days of the hand truck, as many as 20 men were required to unload, weigh, sample, and stack a carload of 100 bales of cotton. Today, with two jeeps and six men the work can be done as quickly as before, and at one-third the labor cost.

Moving into the cardroom, the first application of the conveyor is seen. In the most up-to-date applications, picker laps are distributed to the carding operation by a conveyor system. The old days of doffing, weighing, and racking laps from the picker, followed by a second handling when laps are trucked to the cards and hung, are gone. Now, the lap is doffed onto a conveyor where it is automatically weighed, accepted or rejected, and moved to the carding area where it can be automatically laid up on cards. The result has been a saving in labor and trucking costs which soon pays for such an installation.

Conveyorized handling of materials in processing becomes progressively more important as stock moves through the production operations. The added value of the inventory makes mechanical application of handling or trucking more economical.

Spooling and warping, in particular, readily lend themselves to the mechanical methods of feeding and removing stock. Loading the highly automatic Barber-Colman (BC) spooler from conveyors has increased service labor capabilities tremendously.

Hoists and powered conveyors have encouraged larger packaging at the warpers, with a minimum of labor required for removal and transfer of the stock.

Application of the conveyor principle has been of great benefit in the tire cord industry twisting operations. Crane type conveyors and jeeps have reduced frame loading crews from 50 to 75 per cent. Such labor economies make for lighter work with larger packages, and a reasonable recovery of investment.

In weaving and quilling operations the textile mills have applied materials handling principles to their most effective operation to a considerable extent. Conveyor handling of quills has reduced labor costs for the quilling operation by 80 per cent. One operator can now produce more and better filling at less cost than five operators using the old style equipment.

Warp changers now use hoists to handle large and heavy warps, accomplishing with materials handling equipment twice the man-hour production formerly possible. Hoist conveyors have made possible the processing of larger rolls of fabric by replacing dollies to transport fabric from weaving to finishing. Where two men were used before, only one is required now.

In packing and shipping, the use of conveyors, hoists, and jeeps has simplified and expedited the work, substituting mechanical handling and transport for slower, more expensive manpower.

This section has dealt with some of the major applications of materials handling equipment. In individual situations, such uses as gravity feeds for returning empty bobbins and spools may be applied, or special problems may result in special applications. One such application combined the feeding and loading problems on two operations. The result was a 10 per cent reduction in labor and the recovery of a $48,000 investment in less than 2 years.

Opening and Picking

Machines to blend, clean, and prepare cotton for processing are being developed to a high degree of proficiency. The trend is toward improved basic equipment with regard to cleaning, blending, and increased production. There is no standard approach to the amount or type of equipment used or desired. The quality of the raw stock and the degree of blending desired determine the type and quantity of cleaning and picking equipment required.

The bale breaker and two-process pickers with a man-hour production capability of 250 to 300 pounds, have been replaced by blenders, feed tables, openers, cleaners, distributors, and one-process pickers whose man-hour production capability is 900 to 1100 pounds.

Pickers with pneumatic rack controls and high compression calendar rolls have made heavy laps practical. In some instances the old 50 pound standard lap has been doubled. This increase in lap size, along with better blending and improved handling, is consistent with the new developments in the carding operation which are described in the following section.

Improved opening and cleaning have resulted in savings in areas other than labor. Specifically, when modern equipment is used, a low grade raw stock is capable of yielding an end product equal in quality to that obtained with older machinery and more expensive raw materials. Alternatively, the grade of the raw stock may be maintained to give a product whose quality is so improved that it will command a greater share of the market.

Increased lap weights yield reduced waste in handling and less lap changing at the cards and the reduction in reprocessed fibers further adds to quality. Large laps have also been found to have greater uniformity as compared with the smaller laps formerly produced. Thus, with only one-third of the original labor force and the use of modern machinery, the cost of cleaning and blending has been reduced by as much as 80 per cent and the quality of the product has been augmented.

Carding

As was noted previously, the improved methods of opening and cleaning have fostered and encouraged the development of better carding techniques. However, one major development in carding, namely, vacuum stripping, has resulted independently of these changes.

Normally each card requires one to three strippings in an 8 hour shift, depending upon the stock being processed. For anyone who has stripped cards with a brush, the advantages of vacuum stripping can be seen readily—two men can now do the work that required six or eight men with a brush. Not only is it faster and cheaper to strip by vacuum, but working conditions are greatly improved and more uniform production is possible.

With heavier laps have come conveyors to replace human trucking and hanging on cards. With fewer laps to lay, card-tender assignments can be increased 10 to 30 per cent.

In line with the larger, and hence more economical package, card coilers and cans have been improved. The 12 × 36 inch can has been replaced by the 16 × 42 or even 18 × 42 inch can, allowing further cuts in labor costs. The single coiler has been improved to place 10 to 15 per cent more sliver in a given can. In some instances two strands, or bicoilers, have been added.

Drawing

The operation of drawing, disregarding the larger supply and finished package, is notable for improved speed with reduced variation. Production on today's high-draft 4/5 drawing is in the range of 300 feet per minute as compared to 150 to 180 feet on old style equipment. Also, quality variation has been improved considerably, permitting further improvements in processing in succeeding operations. Automatic doffing and can changing is being introduced.

Roving

The rising young textile man of today will take for granted single process roving, whereas older men remember the short draft slubbers, intermediates, and speeders. Some can even recall a fourth process—the old jack frame. In terms of labor, materials, and floor space, processing roving in this manner was expensive. What has happened?

First, it must be remembered that processing improvements in converting raw cotton to finished fabric are not independent unit developments. Each operation is dependent for its quality and operating characteristics not only on the preceding steps in preparation, but also on how the material is to be handled in subsequent operations. Thus, the vast improvements in opening, blending, carding, and drawing laid the groundwork for reduced roving processes, and the tremendous strides in the utilization of "high" or "super" draft principles in spinning made it practical.

The old processing series, proceeding from slubbers to intermediates to speeders and even to jack frames has, owing to better fiber control and long-draft, been replaced by a single roving frame. No sacrifice in quality has resulted and the savings have been tremendous.

To produce 500,000 pounds a month of 1.85 hank roving on a three-process setup requires three times as many employes as does the same production on single-process roving. The savings in labor are equalled by savings in floor space, power consumption, in-process inventories, and maintenance.

Thus, because of the great strides in the operations before and after the roving process, automation has, to a limited extent, been possible. Three to four processing cycles have been combined into one. Direct sliver to yarn processing is beginning to eliminate the roving frame.

Spinning

Spinning has always been a small package, low production, high cost operation, upon the success of which has rested much of the manufacturer's ability to produce quality goods for market. Any substantial contribution to this operation in the way of automation, with no sacrifice in quality, has been needed for years.

As we have seen, the improvements in each process have led one into another through the entire process. Perhaps the greatest incentive to these developments has been the advances made in high-draft or super-draft spinning. Without the improvements prior to spinning, the tremendous strides in that operation would be seriously curtailed, and in some cases nonexistent.

Today's equipment for converting roving or sliver into yarn is far superior to that in use 15 or 20 years ago. To achieve a 40 to 50 per cent reduction in labor costs, the progressive mill has increased package sizes, increased speeds, and reduced maintenance and cleaning requirements through better design.

Taken in order, both the supply package and the finished package have been increased considerably. Drafts of 15 to 20 have been replaced by drafting ranges up to 70, making possible such changes as larger spinning rings and traverses.

Concurrent with the larger packages are higher speeds. The day when travelers were limited to "mile-a-minute" speeds is gone. Today, spindle speeds of 12,000 to 15,000 rpm and traveler speeds of 7,800 fpm or better have become entirely feasible. No longer is spinning production tied to low speeds and low productivity.

Design changes and auxiliary equipment have aided greatly in reducing costs. "Open" construction of creels and frames has fostered cleaning attachments that eliminate as much as 60 per cent of the cleaning labor. Vacuum collection systems reduce waste, ends down, and lint accumulations. Narrower, more compact frames allow more spindles per bay and reduce floor area requirements. Automatic creeling and doffing are being introduced.

The spinner of today encounters fewer end breaks, reduced creeling and cleaning, and better working arrangements. Therefore, he can tend as many as 80 per cent more sides than was possible a few years ago. Consider also the reductions in service labor (e.g., roving haulers, cleaners, and sweepers) and the extent to which spinning has been "automated" is readily realized.

True automation, however, has not yet come to this operation. The basic principles have not been modified to the extent that feed, processing, and output are a unified and automatic cycle. However, the groundwork is being laid by constant improvements in fiber control and design, and industry is preparing itself for acceptance of total automation.

Spooling and Warping

Automatic spoolers and high speed warpers exemplify the strides taken by the textile manufacturers in utilizing mechanical applications to reduce labor costs and improve handling.

In its simplest applications, automatic spooling has meant savings of from 40 to 50 per cent in processing yarn from warp bobbin to beam. With conveyor servicing of yarn supply, and the larger yarn packages themselves, it is possible to further increase these savings.

Let us look at this machine which has done so much toward automating the process of transferring yarn from a small package to a beam. The author was fortunate in being able to observe, over a period of several years, the use of traditional spooling methods alongside automatic spooling. When the automatic spooler was first introduced into an otherwise unprogressive setup, the productive capacity was completely out of proportion to the existing organization. Much effort and thought were necessary to the final success of the operation.

Here, dramatically illustrated was a lesson in the interdependent changes that have occurred in the industry. It is not always possible to pick out a certain desirable improvement or change and expect it to be smoothly integrated into a series of operations without first adapting the dependent processes to the innovation.

It is well-known among mill men that spooling and warping add little to product value or quality. Weak spot and yarn defects are corrected to a limited extent, but basically the operation merely transfers yarn from bobbin to beam through an intermediate package—the spooler cheese. An operation such as this, where no great advance in processing is made, lends itself very readily to concentrated study on ways and means to by-pass, eliminate, or economize the process. Thus, automatic spooling preceded advances in spinning, since a definite and traceable value is not added to the end product. A cheaper and more efficient method of yarn transfer was highly desirable as the cost of the operation was, more or less, a dead expense.

Today, the principles of automatic transfer are being exploited wherever such operations are in use. Winding, quilling, spooling, and related operations are rapidly becoming more fully automated than the really productive operations of converting fiber into fabric. It is possible that in the future transfer will be accomplished, in many cases, as an integral part of the production operation, thereby automating the purely mechanical steps.

The author does not wish to infer, or leave the impression, that the so called "nonproductive" processing operations are entirely undesirable. On the contrary, these very operations have led the way to advances which may not have been possible, or practical, without the basic improvements in the nonproductive cycles. Automatic spooling and warping have certainly fostered many of the improvements in spinning packages that we see today.

Twisting

Few mills, except for those manufacturing specialty items, engage in twisting operations to any great extent. However, it is in this area that one of the major developments in automated textile production has occurred.

Until 1945, tire cord production as a textile operation paralleled production in textile plants throughout the world. The bulk of the tire cord produced was cotton, requiring the same preparation and spinning operations as any other yarn or fabric produced from a bale of cotton.

For several years prior to 1945 and immediately following the postwar adjustment period, the synthetic continuous-filament fibers rayon and nylon, grew in popularity. Improvements in these man-made fibers paved the way for the conversion from cotton to synthetics—beginning as early as 1950 in some instances.

The use of synthetics constituted automation for the manufacturer in the sense that all of the operations preceding twisting were eliminated. The producer of the synthetic yarn delivered the raw material ready for twisting. The immediate gain, as well as some of the problems involved, is readily apparent. The bulk of the labor force engaged in preparing cotton for twisting into tire cord, as well as the investment in machines and buildings, was now surplus—unless converted to other uses. The finished product capacity, as compared with cotton, was not reduced, although labor and equipment were decreased by one-third. Former textile plants were now used for twisting and weaving operations.

This type of automation was created by a change in raw materials. Arguments as to the advantages or disadvantages of the change became academic in the face of the complete shift of this segment of the textile industry to a completely new concept of production.

As a result of this tremendous change, or perhaps only spurred by it, development engineers began to look at twisting operations with a view to further improvement. The conventional method of converting yarn to tire cord calls for a "cable" twist of two or more plies, which consists of one or more strands of yarn. Normally, synthetic tire cord standards called for twisting a single ply of 12 or 13 turns per inch in one direction and then "cabling" two plies by twisting them 12 or 13 turns per inch in the opposite direction. Under conventional systems this procedure required two separate and distinct twisting cycles and one transfer operation, in order to build the small ply package up to an economical package for use on the cable twister.

Steps were immediately taken to improve packaging and feeding. Upstroke twisters and large ring twisters made their appearance, but proved to be only a temporary measure in the face of the question as to why such twisting should require more than one operation. European engineers answered the question by developing the two-for-one twister which accomplishes the desired result in a single operation.

Paralleling the work of European engineers, American firms pursued their own solutions to this problem. At present there are several machines on the market that give a maximum size finished cord package in a single operation. Automation is indeed inherent in such a machine, as it combines cycles and eliminates processes to give a better product. One man can do the work that was performed by two men using conventional production methods.

Weaving

Generally speaking, the conversion of any fiber into finished product involves three steps: preparation, conversion to yarn, and weaving. Weaving itself involves two steps of preparation; the warp and the filling, followed by the weaving process that combines the warp and filling to make the finished fabric.

Weaving is perhaps the first of any textile operation to have experienced true automation. More than 160 years ago, in 1801, Joseph Marie Jacquard developed a punched card machine to control warp ends individually in the weaving operation. Thus, punched card processing is not a new concept in the textile industry.

In recent years the three steps in the weaving process, (filling preparation, warp preparation, and weaving of fabric) have experienced tremendous advances toward automated cycles and high speed operation. Until quite recently, the emphasis has been on high speeds adapted to equipment which is basically unchanged. Today advanced thinking envisions the combining of cycles to achieve a higher degree of automation in addition to higher speeds.

Filling preparation has advanced from the era of low productivity on winders to a highly productive operation that utilizes almost fully automatic equipment. With new equipment, one operator can produce as much as five operators were able to produce 20 years ago.

The advanced thinking referred to above has carried this operation one step further; the Unifil Loom Winder has combined the filling operation with weaving, creating in itself an automated process which eliminates the entire cycle of quill preparation. Even with highly automatic equipment, such as has been named, the preparation of filling for the looms has involved at least one winding process, transportation, placing quills in loom batteries, and cleaning the quills after the filling has been run off in the looms. The Unifil loom attachment combines these three operations, eliminating labor, transportation, and material cost in addition to improving both the quality and quantity of the finished product.

In slashing and warp preparation the trend has also been toward greater speed. The application of electronic controls to regulate tension, sizing, and drying cycles has helped the industry to produce a more uniform warp and a better quality product. Maximum utilization of materials handling equipment and electrical controls has resulted in a 30 to 60 per cent reduction in labor. Electrical circuits operating on cheap power provide a more positive control than is possible with high priced labor.

Weaving itself has not progressed into new fields of application. Jacquard's application which uses the punched card pattern familiar to all weavers, represents a jump into the future that was not followed up immediately. This was, of course, an application designed to fulfill a specific need, however, the principle could be adapted to other situations. Today, weaving is advanced far beyond the water-powered, hand-change shuttle looms of 100 years ago. Automatic shuttle change, stop motions, and the ability to operate at greatly increased speeds have brought about these improvements.

Administrative Automation

The progress towards increased automation has not been confined to the production cycle in the textile industry. Relatively new in the industry is a growing trend toward the use of electronic equipment in office and administrative work.

The era of electronics began many years ago in this field, with the introduction of calculators, check writing machines, payroll machines, and accounting machines. Data processing equipment is now taking over and opening wider fields. Inventory control and production control, as well as financial and administrative controls, are also becoming automated.

Although computers and data processing machines, are capable of paying for themselves in savings, their greatest value lies in the ability to assemble vital facts quickly and completely. In the complex business world constant and close control is paramount.

The field for data processing is virtually unlimited. When the methods and procedures of today are adapted to electronic processing, tomorrow's problems are created. These in turn can be handled by data processing. Eventually, many administrative and staff functions will lend themselves to this new method, so that receiving, materials, supply, accounting, allocations, billing, and marketing will be handled quickly and accurately from the accumulation of available facts, and at a lower unit cost than is presently possible.

The Future Outlook

What the future holds can only be surmised. Undoubtedly the development of nonwoven fabrics will play a large part in determining how fast and how far this operation will go in improving machinery and methods. High speeds alone are not the answer as there is a limit that cannot be surpassed. Because of present labor costs any operation with the relatively low production per man hour of weaving, as compared with other operations, must be subjected to intensive development of new and better ways to convert yarn into fabric.

The industry is looking for an answer. It may be found in shuttleless looms or in new weaving techniques. Electronics, which plays an ever increasing role in this operation, may be the means of realizing the ideal situation. Certainly no one currently engaged in textile production doubts that this vital operation will continue to experience the advances of the past 20 years at an increasing rate.

The basic steps toward automation in its true sense are being taken by the textile industry, and the economic limitations will be overcome. When it is economically feasible, the textile industry can move into the field of automation, taking full advantage of its progress to date. Plans for a pilot mill, operated on the most up-to-date automation principles, were announced recently.

Historically one of the oldest mechanized industries to come out of the industrial revolution, textiles have been saddled with restrictions not found in other industries of the "electronic age." Once the shackles have been struck, the textile industry can, and should, again become a leader in the development and application of progressive technology.

CHAPTER 13

Quality Control*

Introduction

A product noted for its quality is not produced by accident. It is the result of careful planning and control and good over-all management. Quality, like cost, is thus a measure of the success of management operations as a whole.

A sound mill quality control program assures that for a given product cost, and a relative minimum of control costs, a yarn or fabric of optimum attainable quality is produced. A product which becomes known for its consistently superior quality/cost ratio thus becomes a more salable product.

The means of attaining these objectives are outlined below, after examining the essential aspects of modern quality control, as developed in recent years (1,2).

What Makes for Good Quality

Three principal factors are usually the determinants of good quality. First there is the *design*. A design which is adequate, useful, workable, and attractive is usually the result not only of creative imagination, but also extensive study, planning and research.

Quality is also an expression of *workmanship*, reflecting the pride and skill which go into each processing operation. It is directly related to the quality-mindedness which pervades the plant. Where incentive systems incorporate quality considerations, and employee selection, placement, and training are

* This chapter is adapted from a paper by N. L. Enrick, in *Textile Industries*, **123,** 70–72 (1959). (Reproduced by permission.)

based on carefully worked out programs, mill management's job of instilling quality-mindedness is greatly facilitated.

A third aspect of quality, *uniformity*, has become appreciated only in recent years. Perfect uniformity is generally unattainable, because of variations in raw materials, humidity, machine wear and vibrations, and similar factors. These variations can be controlled within normally allowable limits.

Other deviations of product may be due to some definite cause or "trouble" in the manufacturing process. It is in the maintenance of uniformity where modern quality control becomes important.

The three factors affecting quality of product are summarized in Table 13-1.

TABLE 13-1
Quality as a Function of Design, Workmanship, and Uniformity

Design	Workmanship	Uniformity
	Quality problem	
How adequate, useful, functional and attractive is the design?	Does a sense of quality-mindedness pervade the entire organization?	Is product variability controlled within economically feasible bounds?
	Tools of attainment	
Fabric development Laboratory tests Field tests Customer surveys Market evaluation Market research	Employe selection and training. Quality incentives Standardized processing Follow-up on complaints Quick correction of poor quality.	Analysis and reduction of process variability. Statistical sampling, testing and control. Thorough tests and inspections of outgoing product. Re-check grading.
	Goal	
Useful and appealing styles	High quality, low defects	Uniform, reliable product

Statistical Quality Control

In order to maintain the closest control practically feasible over product variations, a modern quality control program provides tests and inspections at strategic points in manufacturing, from raw material to finished product. Insignificant or normally expected fluctuations may be ignored, but test results showing unsatisfactory conditions or "trouble" are brought immediately to the attention of the operator, his supervisor, or both. In this way, defective output is caught quickly and can be corrected with a minimum loss of time.

Statistical quality control has earned itself a special place in this work, by providing sampling plans which keep the risk of sampling error to a minimum. In addition, control charts may be employed, which are often capable of predicting that a process is about to go "out-of-control" before an actual off-standard product has been produced (2,3).

Quality and Costs

Modern quality controls will not only improve product quality, but also aid in reducing processing costs. This fact becomes apparent when one considers that with controlled production there will be less off-standard output, less waste, fewer ends-down and loom stops, and fewer short-ends and seconds to sell. Thus the axiom that "higher quality means higher costs" can be proved untrue in the mill which keeps proper quality controls.

Given identical raw materials and equipment, the manufacturer who has these controls in his processing is likely to end up with a product of higher quality and lower costs than the competitor who neglects controls and leaves costs, machine efficiencies, and quality to "work out by themselves."

Although quality control in a mill improves quality, a distinction must nevertheless be made between an *improvement program* and a *control program*. Often management will invest in new machinery and equipment to attain higher productivity and improved quality. However, unless such improvement is tied with the proper controls, quality will soon begin to drift, particularly as regards hidden faults, and thus cause a tapering off of consumer acceptance. Thus the salability of a mill's product is closely associated with its quality control program.

Sometimes a mill faced with a sudden demand for higher quality will spend considerable effort on finished cloth grading, often doubling the number of grading inspectors. In this way, the mill's control system will act as a "traffic cop" by which inferior merchandise is sorted out and diverted to "less demanding customers." Such an approach seems to be a costly solution to the quality problem, since *quality cannot be burled or mended into the cloth, but must be spun into the yarn and woven into the cloth.*

Attainment of Good Quality Controls

A good quality control program, based on the principles presented above, must consider raw materials, manufacturing requirements, and market demands. A step-by-step systematic approach towards such a control program, as proved in a large number of mill applications, is outlined below:

1. Initially, the mill should check its records to see that a complete inventory of machinery and equipment is available, with up-to-date information on such machine characteristics as drafting systems, gearing, and constants. Next, data should be sought to show in tabular form the various change gears, speeds, package sizes, and other processing characteristics applicable as different raw stocks, yarns, and fabrics are passed through each machine.

2. Once management has before it, in neat and compact manner, the pertinent data on present machinery inventory and processing setups, possible means of improvement often become apparent. First of all, there may be certain minor changes of processing conditions which may lead to increased machine efficiency, higher production, reduced labor cost, and improved quality. Furthermore, there may be other changes, often involving expenditures of a considerable nature or involving long-term adjustments, which should be kept in mind for a long-range program.

3. It is well for the findings from Step 2 to be put in writing, possibly in the form of an "Attainment Program" listing:

a. Present conditions.

b. Proposed immediate improvement programs.

c. Possible future "long-range" programs.

d. Work required, parts needed, and costs.

e. Estimated cost savings in production versus cost of the program.

4. After the "Attainment Program" has had a thorough review by all parties concerned, it is ready for exhaustive discussion and consideration of the pros and cons by mill management and supervision. As final decisions are reached a record should be kept of each major item, showing:

a. Date item was approved.

b. Date requisite parts were ordered.

c. Date parts were received.

d. Date work was begun.

e. Date work was completed.

f. Results of check-ups on actual improvements attained.

5. Concurrently with these steps, a routine testing program may be initiated, along the lines shown in Table 13-2, and using standardized test procedures (4). After sufficient tests have been accumulated, regular testing schedules can be set up to check all machinery at definite cycles. The testing frequencies, as well as the standards and tolerances, are usually developed by the supervisor of the laboratory in consultation with mill management and supervision. Usually, agreement on what are practical and workable standards, can be attained quickly where all facts and data on machinery, processing, and prior test results are available for review.

6. Next, an off-standard procedure is put into effect. This incorporates the modern principle of "management by exception." This means that so long as test results are satisfactory, no report is made. However, if an actual test exceeds the allowable tolerance, and a re-test and re-check has ruled out sampling error or testing faults, an off-standard report is issued by the laboratory supervisor. This report shows the off-standard condition found, and goes to the departmental supervisor concerned, so that he can correct the trouble.

Only in those cases where the supervisor is unable to take care of the off-standard condition, or where the condition may be

TABLE 13-2
Over-all Design of Yarn Mill Quality Control

Category and purpose of testing	Cotton system (cotton, staple)	Filament system (synthetics)	Woolen system (wool, synthetic blends)	Worsted system (wool, synthetic blends)
Raw stock—check quality of the raw stock and allocate to the most suitable yarns and fabrics.	Length Grade Strength Fineness Maturity Damage Variations	Denier Condition Strength Filaments Dye index Variations	Length Grade Crimp Fineness Variations	Length Grade Crimp Fineness Oil content Neps Variations
Machine speeds—assure that machine speeds stay set for best quality and production throughout the mill.	Opening through Spinning Winding Twisting	Winding Twisting	Dusting through Spinning Winding Twisting	Drawing through Spinning Winding Twisting
Machine settings—check that all machine settings are maintained at standard so as to yield optimum performance.	Beaters Combs Rolls Pins Traverses Spacings	Tensions Traverses Spacings Alignments Rolls	Tapes Tensions Rolls Pans Alignments	Pins Combs Rolls Traverses Spacings Alignments

Other checks—
Sizing: Maintain standard yarn weights.
Evenness: Assure even yarn, free from periodicities.
Run-out: Minimize defects from vibrating (runout) rolls.
Packages: Maintain standard package sizes and handling.
Ends-down: Check on running conditions of stock.
Production: Control efficiency of production.
Waste: Minimize avoidable waste of spinnable fibers.

beyond his immediate control, is the off-standard report permitted to come to higher management's attention. This system avoids any suspicion of "snooping" and keeps down the paperwork and reports.

7. Working along control principles similar to those for quality, a method may be developed also for waste control, incorporating:

a. Proper weighing and classifying of departmental waste.

b. Showing waste in percentage by category and department.

c. Comparing actual waste against waste standards and tolerances.

8. Definite schedules may also be set up for preventive maintenance of machinery. Here again, an off-standard procedure can control the performance of overhaul at the frequencies scheduled, and report substandard quality scouring or overhaul.

9. As these controls begin to function, the mill's finished cloth grading may be reviewed. With streamlined procedures and reports, it should permit management to evaluate at all times the level of quality prevailing for its various styles of fabric and to adjust standards for "first," "second," and "third" grade merchandise quickly in accordance with changing market demands. From an internal viewpoint, the grading records serve as important information for analyzing the types of defects that are occurring and taking ordinary and special corrective actions in the processing departments responsible.

Conclusion

From the facts presented, the gains from a well-run quality control program are apparent:

First of all, there will be improved consumer acceptance and marketability.

Second, there will be reduced waste and seconds.

Third, there will be a smoother flow of work and improved morale, as ends down and loom stops are reduced.

Fourth, a more uniform and controlled output will standardize working conditions and facilitate the application of production incentives.

Quality control, if incorporated in a good mill program,

thus contributes to lowered cost and a more salable product—the best salesmen of goods are the goods themselves.

References

1. Enrick, N. L., *Quality Control and Reliability*, 7th ed., Industrial Press, New York, 1977.
2. Enrick, N. L. *Management Control Manual for the Textile Industry*, New York, Rayon Publ. Corp. 1964.
3. Juran, J. M., Quality Control Handbook, 3rd ed., New York, McGraw-Hill, 1974.

CHAPTER 14

Preventive Maintenance Control

Introduction

Frequently, the industrial engineering or standards department in a mill may be charged with the control of maintenance operations. Proper preventive maintenance of machinery is an important element in efficient and low-cost manufacturing operations, insuring such advantages as longer life of machinery, reduced machine downtime by preventing breakdowns, and generally improved processing conditions. Actual maintenance work is usually performed by the mill's shop or maintenance department. However, the establishment of maintenance frequencies and the control of proper performance often rests with the standards department or some other similar department in the mill.

Control Program

The control program involves the following typical steps:

1. Desirable frequencies of maintenance overhaul of machinery are established, based on the consensus of the superintendent, maintenance foreman, and processing foreman concerned. A typical frequency schedule as shown in Table 14-1.

2. Maintenance check lists are prepared for all types of machinery in the mill, as illustrated by Table 14-2. More detailed lists may be prepared by using suitable recording forms (see Figure 14-1).

3. A record of the maintenance work performed on all machinery is kept in the standards department.

TABLE 14-1
Maintenance Frequency Schedule in Spinning

Type of maintenance work required by spinning department	Maintenance frequency, weeks	Maintenance time required	
		Hours per frame	Hours per week
Overhaul 30 frames	15	8	16
Plumb frame	1	4	4
Rebuff and change top rolls on 30 frames	10	1	3
Overhaul 3 winders	12	4	16
Overhaul 5 twisters	10	5	10
Total	48	22	49

TABLE 14-2
Maintenance Work Checklist on Pin Drafters

Machine part inspected	Machine condition checked
Stock guides	Smoothness and proper positioning
Rolls	Smoothness, less than 0.01 inch run-out
Belts	Wear
Fallers	Pin sharpness, missing pins
Fallerscrews and drive	Good condition
Cams	Timing and setting
Cleaners	Condition and position
Stop motions	Proper sensitivity
Covers	Good fit
Gears	Lubrication, run-out, broken or damaged teeth
Yardage counters	Accuracy
Lubrication system	Proper operation, free from leakages
Panel control	Operating

Machinery Maintenance Control Records

The machinery maintenance control record, usually maintained in a visible records file, forms the central part of the system under discussion. A typical form is shown in Figure 14-2. The following information is revealed:

1. *A maintenance work record* showing the date maintenance work was assigned and the date work was completed on a particular machine.

LOOM SETTING STANDARDS					
MILL	LOOM TYPE			DATE	
CHECK POINT	STD.	TOLERANCE		DESCRIPTION OF MEASUREMENT	
		MAX.	MIN.		

Figure 14-1. Loom setting standards chart.

2. *Major repairs and replacements* which is used as a record in guiding future action on equipment requiring excessive repairs and costly replacements.

3. *Off-standard test results* that provide a guide to the particular machine characteristics requiring close attention. The off-standard test results, be it uneven stock, excessive end breaks, low yarn strength, or other quality items, are posted from the quality control test records.

4. *A monthly inspection control* showing the week and month in which the next maintenance is due for a particular machine.

Coordinated Control Program

From an examination of Figure 14-2 it is apparent that the system described is designed to achieve a coordinated, over-all control program. In particular, the following objectives are accomplished.

1. Maintenance work is scheduled regularly and important items to watch (as shown by the off-standard quality records or excessive replacement records) are called to the attention of the maintenance crew. Thus, maintenance and quality control are connected. In addition, replacement costs are brought under surveillance.

2. The status and condition of machinery are readily known when the following industrial engineering action becomes

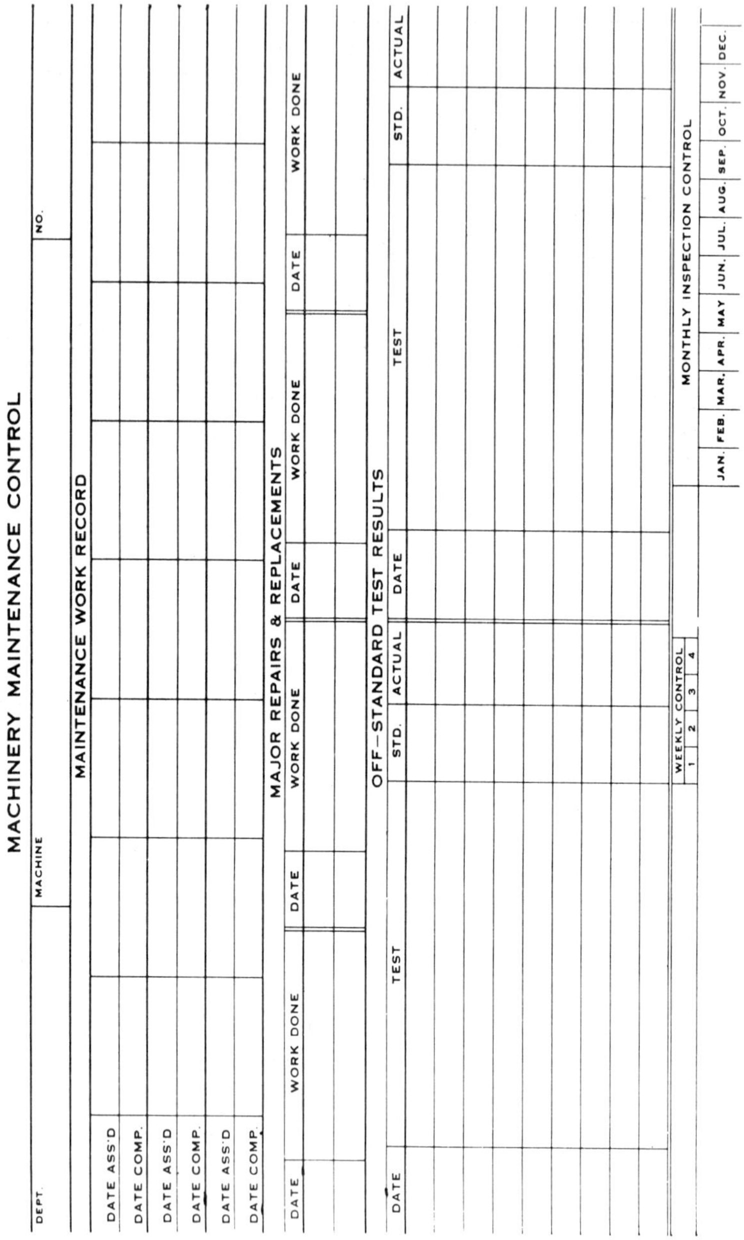

Figure 14-2. Visible record card for machinery maintenance control.

necessary: (a) time study work, (b) review of off-standard running conditions, (c) review of off-standard machine efficiencies, or (d) review of off-standard earnings reports.

3. Maintenance schedules, quality specifications, and cost estimates can be revised on the basis of readily available factual information on actual machine performance and pertinent maintenance work.

Thus, the over-all industrial engineering program benefits from a properly designed and coordinated system of maintenance control.

Maintenance of Uniform Processing Conditions

A good check on the over-all effectiveness of the maintenance program is provided by the degree of uniformity exhibited by the stock produced. Yet, despite thorough maintenance programs, it sometimes happens that stock exhibits excessive variability. In such a case, it may be that some factor in maintenance work may have been overlooked on one or more types of equipment. When excessive variability in stock is encountered, therefore, a review of the check lists provided in this chapter may be useful in locating conditions that may have been overlooked by the routine maintenance work. The check lists, from pickers through spinning are compiled from over 10 years of experience, but may, nevertheless, not represent a complete set of all the maintenance problems that may be encountered in practical mill operations.

In preparing the lists, a .distinction was made between three types of variation, namely, (1) within-delivery variation, (2) within-machine variation, and (3) departmental over-all variation.

In a machine that has only one delivery such as a picker or a card, the within-delivery variation is identical to the within-machine variation. However, in the instance of a machine that has several deliveries, the within-frame variation consists of both within-delivery plus between-delivery variation. Similarly, the over-all variation in a processing department is composed

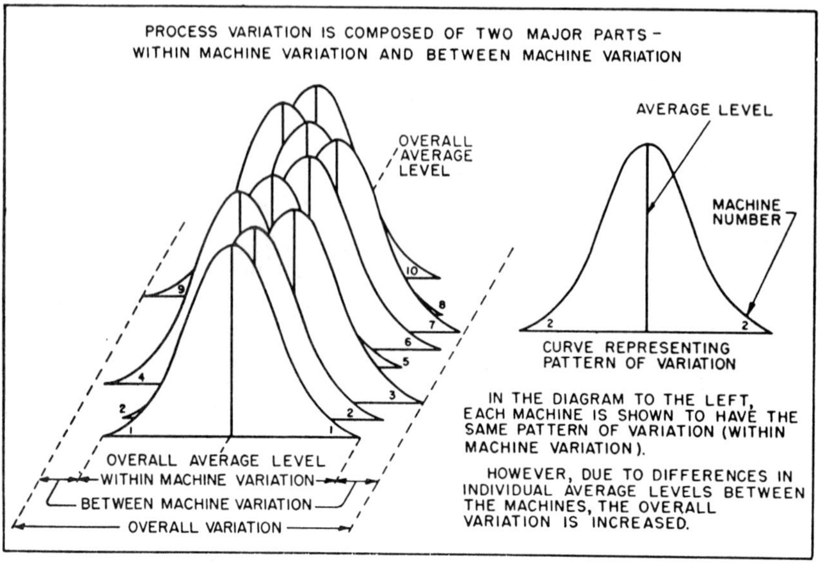

Figure 14-3. Process variation is composed of two major parts, within-machine variation and between-machine variation. (From N. L. Enrick, *Quality Control*, 3rd ed., Industrial Press, New York, 1959. Reproduced by permission.)

of within-machine plus between-machine variation. The manner in which within-machine variation and between-machine variation may combine to yield departmental over-all variation is illustrated in Figure 14-3.

CHECK LISTS

1. Causes of Excessive Variation in Picking

A. Within-lap variation.
  1. Opening line feed is irregular, or stock is improperly opened (clumpy).
  2. Blending reserve level varies excessively, driving belt slips, or plate setting is inaccurate.
  3. Out-of-roundness of calender rolls and gears, fluted rolls, beaters and beater bearings, screens and screen bearings, or loggerhead bearings.

4. Lap pins bent and vibrating.
5. Evener motion insensitive due to belt slippage, lint or grease clog-up, or excessive play in mechanical linkages.
6. Fluted roll wear or wear of fluted roll drives.
7. Screen surface uneven or clogged.
8. Fan speeds improper causing uneven, excessive, or inadequate air passages, such as turbulent air flow at screens—excessive air pressure in waste box.
9. Improperly set or choked dampers.

B. Lap-to-lap variation on picker—excessive variation in lap weight, from lap-to-lap on the same picker, may be due to (1) excessive or inadequate weight adjustments by the operator; or (2) unsatisfactory lap weighing scales that do not give fine enough readings to permit proper picker adjustments based on actual lap weights. Lack of instruments showing moisture content and providing for compensating devices, such as moisture-regain counter-weights for lap scales, may cause excess variation in true weight from lap-to-lap. A variation in lap length causes faulty weight values.

C. Over-all variation—excessive over-all variation indicates differences in actual draft between pickers. This resultant variation in lap weights between pickers, added to the lap-to-lap variation within pickers, and the variation within laps, yields over-all variation. The latter is fed into the cards and drawing.

2. Causes of Excessive Size Variation in Carding

A. Excessive within-card variation.
1. Worn doffer or feed roll bearings; unbalanced doffer.
2. Untrue calender, lap, or feed rolls.
3. Improper licker-in speeds.
4. Undue slippage of lap from lap roll to feed roll.
5. Defective or chattering gears in bonnet, or chipped production gears.
6. Unsatisfactory clothing, improper surface grinding, or slack chains.
7. Excessive variation in the lap or bent lap pins.

B. Excessive over-all variation—indicates differences in actual draft between cards, caused by one or more of the following:
1. A difference among lap rolls on different cards ($^1/_8$ inch or more).
2. Differences in settings among card grinders' sections.
3. Wrong gears in drafting train, particularly tension gears.
4. Differences in age and wear among card sections, thereby demanding differences in tightness of settings.
5. Differences in trumpet sizes (especially if cast iron).
6. Differences in average lap weight off different pickers which are then transferred into the cards.

3. Causes of Excessive Size Variation in Drawing

A. Excessive within-delivery or within-frame variation.
1. Weight hooks elongated, bent, and rubbing the steel rolls.
2. Adjustable weight hangers too long and resting on the weight-release bars; weights off.
3. Weight linkage binding and weights resting on bolt ends.
4. Worn or broken roll stands.
5. Scored roll necks.
6. Bent rolls (due to driving gears off and on); roll run-out.
7. Damaged gears (chattering or broken teeth).
8. Flattened top rolls with worn coverings.
9. Gears not meshed properly, or backlashing.

B. Excessive over-all variation—indicates differences in actual draft between frames, caused by one or more of the following:
1. Variation transferred from card groups that run at different average level and feed in the channeling (rather than random) flow of drawing.
2. Different draft or crown gears in use on some stock.
3. Differences in trumpet sizes among different frames. (Note: If different trumpets are used on one frame, excess within-frame variation results.)
4. Differences in tension gears among different frames.
5. Differences in actual draft among different make, year, or wear of frames, not compensated by gearing.
6. Differences in weighting used on frames.

4. Causes of Excessive Size Variation in Combing

A. Within-delivery or within-machine variation.
1. Nonstandard drafts, such as between lap roll and feed roll or between detaching roll and calendar roll.
2. Rolls, eccentric or sticking.
3. Nipper knives bent or misaligned.
4. Needles broken or bent; improper needle spacing.
5. Timing of top combs or detaching rolls off-standard.
6. Gears damaged (chattering or broken teeth), or not meshed properly.
7. Roll stands worn or broken; roll necks scored.
8. Weight linkage binding, or elongated and bent weight hooks.
9. Aspirator system choked.

B. Over-all variation—excessive over-all variation indicates differences in waste extraction or actual draft between frames, caused by one or more of the following:
1. Variation transferred from lap-winder or ribbon lappers run at different average level, and feeding in channeled (rather than random) flow to combing.

2. Differences in draft or tension gears, or trumpet sizes.
3. Differences in noils removed.

5. Causes of Excessive Size Variation in Roving

A. Excessive within-delivery or within-frame variation.
   1. Builder motion belt slippage.
   2. Roll defects: fluted—marks caused by the flutes of steel rolls; channelled—a groove is formed by constantly running the stock over the same surface; flattened—by leaving the weights on while not running; rough surface—due to uneven wear or loosening of some of the surface of the cover; loosened cots—due to adhesive releasing the cloth, leather, cork, or synthetic, or to leather stretching; variable diameters—due to poor cloth or leather or to poor buffing, or uneven wear on the frame.
   3. Torsional roll vibration.
   4. Clogged or chipped flutes or condensers.
   5. Faulty gearing; worn, damaged or improperly meshed gears.
   6. Worn roll bearings or scored journals.
   7. Stretching of sliver fed due to uneven tension.
   8. Differences in twist in sliver fed.
   9. Differences in roll settings within frame.
   10. Faulty weighting device or weights not properly applied.
   11. Worn or broken roll stand; waste collected.in roll stand.
   12. Variations in bobbin diameter.
   13. Bent flyers or rough spots on flyers.
   14. Differences in tensions between spindles.
   15. Irregular or clogged aprons.
   16. Slack chains.
   17. Excessive bobbin diameter variation (I.D. and O.D.).

B. Excessive over-all variation.
   1. Variation transferred from drawing through channeling rather than random flow of stock.
   2. Use of different gears (lay, tension, draft, or twist) between frames on the same stock.
   3. Differences of roll diameters or roll settings between frames.
   4. Differences in models and makes of frames.
   5. Differences in roll coverings between frames.
   6. Differences in weighting between frames.
   7. Differences in draft distribution between frames.

6. Causes of Excessive Size Variation in Spinning

A. Within-delivery variation.
   1. Skewers, skewer bearings, or bobbin suspension holders defective or clogged with waste.

2. Stretched roving from roving frame, misplaced roving bars, or faulty roving trumpets.
3. Rolls worn, run-out misaligned, improperly set or weighted, or improperly covered (static-producing).
4. Drafting aprons worn, grooved, seamy, or slipping.
5. Bearings worn, misaligned.
6. Gears defective or not meshing properly.
7. Lifter rods worn or sliding.
8. Spindles improperly lubricated, running on worn bearings, bent or otherwise out of balance.
9. Rings worn or out of alignment with spindles.
10. Travelers worn or of improper weight.

B. Within-machine variation.
1. Differences in roving twist, tension, or size from different frames.
2. Differences in trumpet sizes.
3. Unequal tape tensions and slippage.
4. Cylinder or bearings misaligned.
5. Differences in roll weightings; or bent, damaged, or otherwise faulty weight hooks or saddles.
6. Differences in gearing between the sides of the frame.
7. Unequal roll or ring wear.

C. Room over-all variation.
1. Differences in actual draft or contraction among frames.
2. Differences in over-all frame conditions, or in model and make.
3. Differences in roll weightings or settings.
4. Differences in roving size from different frames, channeled into spinning frame groups.

PART III

OPERATIONS RESEARCH TECHNIQUES

CHAPTER 15

Machine Interference

Introduction

It is unlikely that any industrial engineer working in the textile industry is not familiar with the phenomenon of machine interference and the problems associated with it; however, let us start at the beginning.

Taking a simple case and ignoring complications for the moment, we may picture a weaver in charge of a group of looms and imagine that at the instant he is observed all his looms are running. His aim is to keep them running and producing as much of the time as possible, but any loom is liable to stop at random. When that happens, he knows what to do. If he is free, he goes at once to the stopped loom and fixes or corrects the cause of the stoppage, and restarts the loom. Again all are running. The production of that loom has had an interruption, such as 30 seconds, which has to be taken into account in any standard set. If the looms are all similar it is assumed that, on the average, they all have the same amount of loss due to such stoppages, so that a standard set for one loom will be valid for all.

The time loss mentioned above is not machine interference. It is a characteristic of the material being manufactured and is independent of the number of looms in the group. Imagine that the next time the stop occurs, the weaver is not so fortunate. Before he can get the stopped loom back into production, another loom stops. The second loom has to wait until the weaver is free to attend to it. How long that will be, will depend on the stage reached by the weaver in his work on the first loom. He may have only just begun, in which case a long wait is in prospect, or he may be almost through, in which case the wait may be almost nil. When he does get around to the second loom, he gives it the necessary attention and restarts it. All the looms are in production again.

The time loss representing waiting time imposed by the overlap of machine stoppages is called machine interference. The inevitable, but relatively constant, attention time can be obtained by time study, but the "chancy" and variable waiting time is more elusive. In the example above it can lie anywhere in the range of 1 to 30 seconds. Now, suppose a third machine had stopped during the weaver's spell of trouble. The range of possible waiting times for the third is greater still, for it may have stopped early in the weaver's work on the first loom, or it may have stopped late in his work on the second. In the latter case it would have almost no wait, and in the former it might wait almost the full duration of two repairs. There is no guarantee that more looms may not stop, and the possibilities are compounded.

The weaver will have his good spells and his bad spells. The standard has to be set so that it is fair, and that means it will have to be based on averages. What is the average amount of interference? This is the classical problem of multimachine assignments. If it could be solved, it would be possible to evaluate the average cycle time for a given fabric style. The cycle time is the time in which a stated unit of production is delivered, and includes the time during which the machine is running, plus the time during which the machine is at rest. The rest time is composed of the waiting time or interference, and the time during which the machine is at rest but the operator is working on it. For this, the abbreviation "marow" will be used.

The above statement of the problem, though it does not cover all the ground, typifies many actual situations. To solve the problem as stated would be a big advance. Later some of the complications could be considered.

Looms have been mentioned as a type of machinery that provides the best known example of interference, however, there are many other machines in the textile industry, and in other industries, which have similar characteristics and are subject to loss from similar causes. These machines all have a running time which is long in relation to the operator attention time, so that they can be assigned in numbers to one operator. Once this is done, overlapping stops can occur in most cases, but not necessarily in all. If the stops are regular they can be scheduled and interference may be avoided. Most machines have some regular stops and some random stops, and the effect is practically the same as if all were random.

For the few machines which have only scheduled stops the term "ideal automatic" is proposed, for the others, the name "semiautomatic" will be used. In the case of the ideal automatic machines it is relatively easy to calculate interference, when it occurs, and it is proposed to deal with this phase first as a stepping stone in overcoming the greater difficulties of the semiautomatic.

The operator, for the purpose of this study, has only one pace of work. He never hurries. He never dawdles. He always works at the same steady purposeful rate. He is a standard operator—strictly a concept of the industrial engineer, who finds it useful to imagine this average man whose work times will be valid for setting standards. Thus, the times of operator tasks become constants just as the machine time is constant. When allowances are included in the task time, the operator is assumed to take exactly the specified amount of time. Where necessary, account will be taken of the effects to be expected if the operator pace is nonstandard.

Ideal Automatic Machines

Many of the concepts introduced here will be of use in later sections. Consider an ideal automatic machine with a run time, T, of 90 seconds. When the machine stops the operator walks to it, unloads it, loads it, restarts it, and then rests. All of this takes 30 seconds, t. The cycle repeats itself. The total cycle time, H, is 120 seconds. The machine has no waiting time. The rest, if included, must be regarded as an essential part of the operator's work time. The operator, however, has a great deal of waiting, in fact, 75 per cent of his time is spent in this manner.

Naturally this situation would not be regarded as normal. Assuming they were available, more machines of the same type, such as certain looms, could be added to the operator's assignment. It is clear that he could deal with three other similar machines while the first loom was making its run. By the time he had started the fourth, the first would have stopped, but, since he would be ready for this stoppage there would be no waiting time. Similarly, by the time he had finished repairing the first loom, the second would be ready for him, and so on. There would be no machine waiting, and there would be no idle time for the operator. He would be fully occupied. His machines would all be at maximum productivity. When steady running conditions have been achieved, each machine is giving one unit every 120 seconds, and the operator is responsible for an output of four units every 120 seconds.

Units of production are myriad such as yards of cloth, pounds of nails, miles of wire, and bobbins filled. The one common denominator for all automatic machines is the machine running hour, mrh. The use of this unit of production simplifies discussion of the problems involved. The efficiency, E, of a machine is measured by the average fraction of an mrh it achieves per hour. The productivity, A, credited to an operator is the total mrh of his machines as a group in an hour, on the average. It is plain that E equals T/H, and that A equals $(nT)/H$, or $(n)(E)$, where n represents the number of similar machines in the care of one operator.

Note, that T/H is the ratio of two time values and, therefore, is dimensionless. It makes no difference if the result is labeled "seconds per second," "hours per hour," etc. Generally, we will use hours per hour. When the operator tends four machines under the conditions outlined, E is the same as if he had only one, namely, E is 0.75. A equals 4×0.75, or 3.0 mrh per hour.

The operator's burden, B, is $(nt)/H$, 1.0, in this case. This means the operator is busy for 60 minutes per hour (relaxation is considered as part of his work). Therefore, he cannot possibly make more than four machines produce at one time, and 3.0 mrh per hour is his maximum productivity. In the ideal situation envisaged, there is no waste of either machine time or operator time. If he were given an extra machine to run, there would inevitably be machine waiting time. For later work, however, it will be instructive to see what happens when a fifth machine is assigned to the weaver.

At the instant loom no. 4 is started, no. 1 stops, however, the operator has to proceed to no. 5 where he spends 30 seconds. It is only at the end of this time when no. 5 loom is running that is he free to return to no. 1. The cycle is then complete. After stopping, the first loom has to wait 30 seconds for the operator, and this time loss is passed along so that each machine, in turn, has to wait 30 seconds. The 30 second wait has become, therefore, part of the cycle time. H, in fact, is now 150 seconds (5×30) instead of 120. If there were six machines, H would be 180, and so on. In general, H is given by nt for assignments in the overload region. On a per cycle basis the operator is getting more production, but, if calculated on a per hour basis, not a mite over the 3.0 he got at an assignment of four looms. Since A is $(nT)/H$ and H is nt, then A equals T/t which is a constant, regardless of the size of n. This is true in the overload region only.

It is obvious that cycle time, when the operator is underloaded, is given by H equals $T + t$. When an overload occurs the cycle time changes from $T + t$ to nt, and the critical assignment, n_c, at which this happens is given by

$$n_c = (T + t)/t = 120/30 = 4$$

A useful parameter is the cycle ratio, p, defined as the duration of the machine rest, operator working, or "marow" per mrh. In other words, p is t/T. Then, n_c equals $(1 + p)/p$ and A_{max} equals $1/p$.

Notice that A_{max} is equal to $n_c - 1$, and the common sense interpretation of this is that when an operator has n_c machines, one is being serviced and $n_c - 1$ are running—there are none waiting.

The machine efficiency is T/H. It remains constant up to assignment n_c, since H is constant in that range. When n exceeds n_c and H starts to grow, then E begins to shrink.

H is increased by the waiting time or interference, and there is no difficulty in calculating it in simple cases, as has been demonstrated. It would make no difference to the calculations if the operator elected to ignore the fifth and subsequent machines, if any. The average machine efficiency would decline exactly as

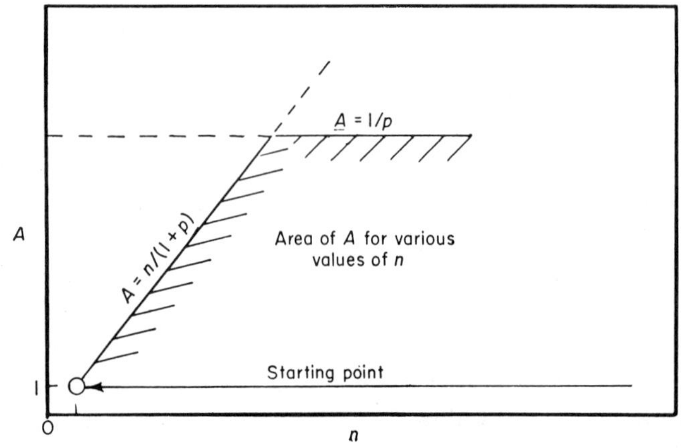

Figure 15-1. Plots of curves of A corresponding to various values of n randomly serviced semiautomatics, fall wholly within the shaded area, beginning with the starting point indicated.

before. Table 15-1 gives in Part A all the results of interest for assignments of one to eight machines per operator. Clearly, it would be folly to assign more than four looms, but the additional values are illuminating. Lines 6, 7, and 8 contain the story of 1 hour in the life of one machine at each assignment. For instance, at n equals 5, each machine will run for 60 per cent of the time and stand for 40 per cent, and half of the standing time is interference. Lines 10 and 11 tell how the operator's average work hour is split. At n is 3, for example, he will work 75 per cent of the time and will be idle 25 per cent.

Since the first example was tailored to present a simple picture, it is natural to ask what sort of a picture is found when T is not a perfect multiple of t, and which is much more likely to be the case. What happens for example when T is 80 and t is 30? Here p equals 30/80 or 0.375, A is $1/p$ or 2.67, and n_c equals $(1 + p)/p$ or 3.67. Part B of Table 15-1 shows the results for this case.

Sometimes it happens that the operator has work to do while the machines are running, in addition to his work on stops. For instance, he may have to inspect and label the product he has just unloaded. Let the time spent on such tasks per machine per cycle be t_1. It is assumed that t_1 is less than T. Since this is so, t_1

TABLE 15-1
Ideal Automatics—Statistics for Various Assignments

	Part A		$T = 90$ and $t = 30$					
1. n	1	2	3	4	5	6	7	8
2. nt	30	60	90	120	150	180	210	240
3. H	120	120	120	120	150	180	210	240
4. $A = (nT)/H$	0.75	1.50	2.25	3.00	3.00	3.00	3.00	3.00
5. $Wm = H - (T + t)$	0	0	0	0	30	60	90	120
6. $E = T/H$	0.75	0.75	0.75	0.75	0.60	0.50	0.43	0.375
7. $Z = t/H$	0.25	0.25	0.25	0.25	0.20	0.17	0.14	0.125
8. $Im = (Wm)/H$	0	0	0	0	0.20	0.33	0.43	0.50
9. $Wo = H - nt$	90	60	30	0	0	0	0	0
10. $Io = (Wo)/H$	0.75	0.50	0.25	0	0	0	0	0
11. $B = (nt)/H$	0.25	0.50	0.75	1.0	1.0	1.0	1.0	1.0
	Part B		$T = 80$ and $t = 30$					
1. n	1	2	3	4	5	6	7	
2. nt	30	60	90	120	150	180	210	
3. H	110	110	110	120	150	180	210	
4. $A = (nT)/H$	0.73	1.45	2.18	2.67	2.67	2.67	2.67	
5. $Wm = H - (T + t)$	0	0	0	10	40	70	100	
6. $E = T/H$	0.73	0.73	0.73	0.67	0.53	0.44	0.38	
7. $Z = t/H$	0.27	0.27	0.27	0.25	0.20	0.17	0.14	
8. $Im = (Wm)/H$	0	0	0	0.08	0.27	0.39	0.48	
9. $Wo = H - nt$	80	50	20	0	0	0	0	
10. $Io = (Wo)/H$	0.73	0.45	0.18	0	0	0	0	
11. $B = (nt)/H$	0.27	0.55	0.82	1.0	1.0	1.0	1.0	

Key: Wm is machine waiting time per cycle; Im is machine wait per hour; Wo is operator waiting time per cycle; Io is operator wait per hour; t is time per cycle operator is at work on one machine (machine stopped); Z is time per hour operator is at work on one machine (machine stopped). Time is given in minutes per cycle and decimal hours per hour.

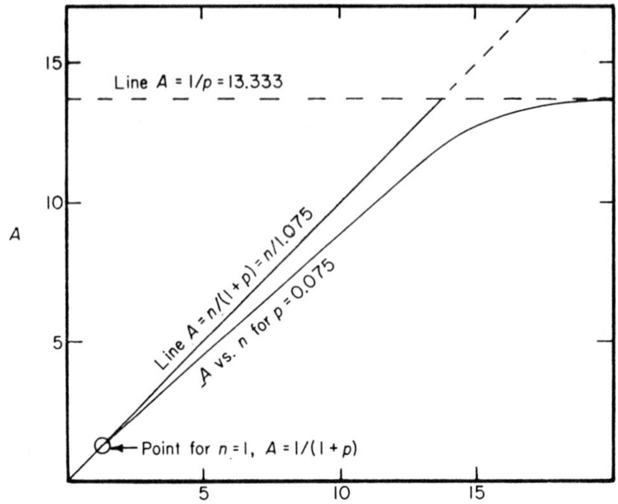

Figure 15-2. Productivity with random stops and random servicing.

does not affect the length of the cycle as long as n is less than n_c, and, therefore, the cycle time remains at $T + t$. When n is greater than n_c, however, it is clear that H must be $n(t + t_1)$. Then, n_c is given by $(T + t)/(t + t_1)$, A_{max} is $T(/(t + t_1)$, and B is $n(t + t_1)/H$. B becomes 1.0 when n is equal to, or greater than, n_c.

Consider a case where T is 90, t is 20, and t_1 is 10, so that n_c is 3.67 and A_{max} is 3.0. The relation $n_c - 1$ equal to A_{max} no longer holds. Table 15-2 shows the results for this case. It is evident from the tables presented that on ideal automatic machines the waiting time, if any, is either machine waiting time or operator waiting time, never both. These machines differ radically from the semiautomatic type, where, if the assignment is greater than 1, there is waiting time for both man and machine.

To select a machine assignment when n_c is not a whole number, let n_L and n_h represent the whole numbers adjacent to n_c. In choosing between n_L and n_h, production cost would be a reasonable criterion. The following example shows how it could be applied:

Suppose the amount of overhead charged to each machine per hour of use is U, in units of an hour's pay. Then the hourly

TABLE 15-2
Ideal Automatics—Statistics for Assignments Using $T = 90$, $t = 20$, and $t_1 = 10$

1. n	1	2	3	4	5	6	7	8
2. $n(t + t_1)$	30	60	90	120	150	180	210	240
3. H	110	110	110	120	150	180	210	240
4. $A = (nT)/H$	0.82	1.64	2.45	3.0	3.0	3.0	3.0	3.0
5. $Wm = H - (T + t)$	0	0	0	10	40	70	100	130
6. $E = T/H$	0.82	0.82	0.82	0.75	0.60	0.50	0.430	0.375
7. $Z = t/H$	0.18	0.18	0.18	0.17	0.13	0.11	0.095	0.083
8. $Im = (Wm)/H$	0	0	0	0.08	0.27	0.39	0.475	0.542
9. $Wo = H - [n(t + t_1)]$	80	50	20	0	0	0	0	0
10. $Io = (Wo)/H$	0.73	0.45	0.18	0	0	0	0	0
11. $B = [n(t + t_1)]/H$	0.27	0.55	0.82	1.0	1.0	1.0	1.0	1.0

cost of the lower assignment n_L will be 1, the operator's pay for the hour plus Un_L, the cost of the n_L machines for the hour. The total is $1 + Un_L$. Production in the hour is, say, A_L mrh. Therefore, the cost of producing 1 mrh at assignment n_L is $(1 + Un_L)/A_L$, which may be designated UC_L. Similarly at the higher assignment n_h, the cost, UC_h, is $(1 + Un_h)/A_h$. When both of these unit costs have been evaluated, select the assignment associated with the lesser of the two. Taking the figures from Table 15-2, and adding the information that U is equal to 0.6 (assumed), UC_L is 1.145 and UC_h is 1.133.

The answer indicates that assignment 4 will give slightly lower operating cost than assignment 3, and would, therefore, be chosen if cost were the dominant consideration. Notice, however, that if productivity were important, with customers demanding delivery, an n of 3 gives a higher productivity than an n of 4, and thus, the issue would be still in debate.

Now for a given set of time study results, there must be a critical value of overhead, U_c, at which the two unit costs are equal, and at which it is immaterial from cost considerations alone which of the two assignments is used. By equating UC_L and UC_h, it is found that

$$U_c = \frac{(T + t) - n_L(t + t_1)}{n_L n_h(t + t_1) - n_L(T + t)} = 0.667$$

If U is less than U_c, it will pay to use the higher assignment$_c$ and vice versa. In the example, U is 0.6 and this is less than U, at 0.667. Therefore, an n of 4 is better than an n of 3 as regards cost. This checks with the above.

Semiautomatic with Cyclic Servicing

When stops occur in regular sequence it is obvious that servicing should be synchronized to conform to this sequence. A definite patrol cycle results, and is designated as "cyclic servicing." When stops are not synchronized or where they occur at random, the job method may call for the operator to always tend the nearest stopped machine, thus representing random servicing. However, even where stops occur at random, cyclic servicing is often preferred and will be discussed here. A presentation of random servicing will be given thereafter.

Cyclic servicing may be performed in an open or closed circuit. In a closed circuit the operator patrols the array of machines assigned to him in such a manner that, upon completing his cycle, he is back at the starting point. In an open circuit, on the other hand, the completion of a patrol places the operator at the far end of the starting point. The discussion below will be in terms of open circuits, but the appropriate modifications for closed circuits will be indicated. It should also be noted that in true cyclic servicing there is no backtracking. Thus, theoretically, if the operator notices a machine stopping just a little behind his present patrol position, the operator will not backtrack to attend to it.

Assuming the operator's pace does not vary, whether he is walking or working, the time taken for a patrol will depend upon the number of machine stoppages in his patrol cycle (assuming also that the average repair times are roughly equal), whereas the number of stoppages encountered will depend to some extent on the duration of the patrol. The shorter the patrol, the less chance of stops occurring, and vice versa. The operator's practice with regard to his relaxation is also a factor. It seems reasonable to stipulate that the standard operator will take a breathing spell at the end of each patrol. If the practice is different among actual operators, it is believed that the final result will not be significantly different. The practice of stopping all machines for a rest period instead of snatching a minute here and there is different in result. It is easy to calculate, but as it is not usual, it is omitted.

Under the condition of cyclic servicing as defined above, it is axiomatic that, once a machine stops, it will remain at rest half of the average time of one patrol. With this fact as a starting point, the machine efficiency may be found thus:

Average time to patrol the n machines, in minutes	y	(1)
Machine running minutes ideally possible per patrol	ny	(2)

Average number of stops encountered per patrol	s	(3)
Average lost time per patrol, machine in minutes	$(sy)/2$	(4)
Average machine minutes running time per patrol	$(ny - sy)/2$	(5)
Machine efficiency, E, line (5)/line (2)	$1/2 - (s/2n)$	(6)

Line (6) states the required expression for the machine efficiency for a given assignment with a standard operator in terms of the average number of stops per patrol (which must be evaluated before the expression can be used). Study of the problem reveals that y can be expressed in two forms, both containing s. On eliminating y by equating those two forms, an expression for s is revealed.

Average number of patrols per clock hour	$60/y$	(7)
Average total stops per hour on n machines	$(60s)y$	(8)
Average number of stops per hour on one machine	$(60s)/ny$	(9)
Average number of stops, q, per mrh (since 1 mrh requires $1/E$ clock hours)	$(60s)/Eny$	(10)
With use of line (6) to eliminate E, q equals	$(120\ s)/y(2n - s)$	(11)
From line (11) by rearrangement, y equals (this is the first expression for y).	$(120s)/q(2n - s)$	(12)
Average operator work time per stop, in minutes,	t	(13)
Time to walk between adjacent machines, in minutes	w	(14)
Walk and work time in average patrol	$2nw + st$	(15)
Standard allowances, leaving k fraction of hour per hour	k	(16)
Therefore, y equals (this is the required second expression)	$(2nw + st)/k$	(17)

Equating (17) and (12)

$$(120s)/q(2n - s) = (2nw + st)/k \tag{18}$$

it follows that

$$qts^2 - [2qn(t - w) - 120k]s - (4qwn^2) = 0 \tag{19}$$

For convenience rewrite (19)

$$as^2 - bs - c = 0 \tag{20}$$

where a is qt, b is $2qn(t - w) - 120k$, and c is $4qwn^2$ then

$$s = [b + (b^2 + 4ac)^{1/2}]/2a \tag{21}$$

The values of q and t will be available from time studies, and n will be assumed for trial if not given as problem data. Then, (21) may be used to obtain s, and (6) to obtain E.

Walk time, w, will be computed as $0.005L$ minutes, where L is the distance in feet from center to center of adjacent machines. This time allowance corresponds to a little less than 3 miles per hour. It is an arbitrary figure and different people will have different ideas as to what is adequate. Later, when considering random servicing, 0.004 minutes per foot will be used (based on the assumption that a man will walk a little more briskly when going to a definite location). If these suppositions are out of line the effect on the total picture will not be significant.

In line (15) walk time per patrol is $2nw$, indicating an open circuit arrangement of machines. If the arrangement is a closed circuit, substitute nw for $2nw$. Then, a equals qt, b is equal to $qn(2t - w) - 120k$, and c is $2qn^2w$.

The value of k is obtained from the standard allowances expressed as a decimal. If the operator is allowed 10 per cent of total time, then k is $1.0 - 0.1$ or 0.9. A 15 per cent allowance would mean k is 0.85. Sometimes allowances are based on the work time—10 per cent of the work time would be different from 10 per cent of total time. To find the decimal of total time, g, when given the decimal of work time, h, use g equals $h/(1 + h)$. For example, if h is 0.10, then g will be 0.091 approximately, and k will be 0.909. If $h = 0.15$ then g is 0.13 and k is 0.87.

To illustrate the use of the formula, it will be applied to the following problem:

Given—an average of 10 stops per mrh (q is 10); an average repair time per stop of 0.3 minutes (t is 0.3); a relaxation of 15 per cent of total time (k is 0.85); a distance from center to center of adjacent machines of 6 feet (L is 6, so w is 0.03); overhead per hour of machine, 20 per cent of an hour's pay, (u is 0.20); and an open circuit.

Find—how many machines should be assigned to one operator for minimum production cost, and the standard efficiency at this assignment?

First, establish the constants:

$$a = qt = 3$$
$$b = [2qn(t - w)] - 120k = (5.4\text{n}) - 102$$
$$c = 4qwn^2 = 1.2\text{n}^2$$

Next, set up a scheme as shown in Table 15-3.

It appears that an assignment of 12 machines answers the requirements of minimum cost, with machine efficiency E equal to A/n which equals 10.21/12 or 0.85 (85 per cent). Symbols d and f are merely stepping stones in the work.

Before interference can be calculated (though it is not important to know the figures, it is at least interesting) it is necessary to find the length of the average patrol, y.

Walk is equivalent to $2nw$ or 0.720 minutes. Work, st, equals 1.074 minutes net. Then y is 1.794/0.85 which equals 2.111 minutes, and 1.055 minutes represents the duration of the average stop. The repair time of each stop is 0.3 minutes. Thus, by difference, the interference is 0.755 minutes, which is almost exactly $2^1/_2$ times the repair time. The average work hour of a machine may be broken down as follows:

Run	0.8500
Marow	0.0425
Interference	0.1075
Total	1.0000

If the machines were of the ideal automatic type, 16 could be assigned without loss of efficiency (numerically equal to 0.94).

Thus, 0.94 − 0.85, or 0.09, is the toll taken by interference. It represents a productivity loss of 9.5 per cent.

TABLE 15-3
Semiautomatics—Cyclic Servicing of Random Stops

1. n	10	11	12	13	14
2. $5.4n$	54	59.4	64.8	70.2	75.6
3. b = line (2) − 102	−48	−42.6	−37.2	−31.8	−26.4
4. b^2	2304	1812	1383	1010	696
5. n^2	100	121	144	169	196
6. $c = 1.2n^2$	120	145.2	172.8	202.8	235.2
7. $4ac = 12 \times$ line (6)	1440	1742.4	2073.6	2433.6	2822.4
8. d = lines (4) + (7)	3744	3554	3457	3444	3518
9. $\sqrt{d}$	61	59.6	58.7	58.6	59.3
10. f = lines (3) + (9)	13	17	21.5	26.8	32.9
11. $s = f/2a$	2.167	2.83	3.58	4.47	5.48
12. $s/2$	1.083	1.415	1.79	2.235	2.74
13. $A = n - s/2$	8.917	9.585	10.21	10.765	11.26
14. Machine cost = Un	2.0	2.2	2.4	2.6	2.8
15. Total cost	3.0	3.2	3.4	3.6	3.8
16. Unit cost[a]	0.3365	0.334	0.333	0.3345	0.338

[a] These costs are given in terms of the operator's hourly pay. A cost of 3.2 units is equal to the operator's pay for 3.2 hours. At a pay rate of $2.00 per hour, the total pay would be equal to 3.2 × 2 or $6.40.

Semiautomatic with Random Servicing

In dealing with machines which have only scheduled stops, machine efficiency can be found very simply, as has been seen. When the stops occur at random, it has been shown that the solution is only a little more difficult, so long as the operator services the machines on a cyclic routine. In the case now to be considered, however, where both stops and servicing are random, the use of the mathematics of probability is necessary.

For those who are not at home in this region of mathematics, a good deal of faith in the mathematician is necessary. Even if it is not possible to meet him on his own ground, there is always the test of common sense that can be applied to the results without understanding exactly how the results were obtained. In the present instance, common sense can indicate the boundaries of the area wherein the true answer must lie. If the mathematician's answer does not fall into this area, it cannot be correct; however, if it does and if all the answers derived from it are self-consistent, then it merits confidence until proved wrong.

It was previously shown that the productivity of n ideal automatics in the care of one operator will follow these two simple rules:

(1) A equals $n/(1 + p)$, from n equal to 0 to n equal to $(1 + p)/p$, and (2) A is equal to $1/p$ (therefore constant) for n equal to $(1 + p)/p$ and all higher values of n. Common sense can now derive the following for the case of random stops and random servicing:

1. At n equal to 1, there can be no interference, so A for semiautomatics must be the same as for the ideal automatic, namely, $1/(1 + p)$.

2. For any given higher value of n, A for the semiautomatic must always be less than A for the ideal, since the automatic is subject to interference.

3. Draw a figure of two axes and label the vertical axis A and horizontal axis n. Draw in a horizontal line to represent A equal to $1/p$, and a sloping line starting from the origin to represent A equal to $n/(1 + p)$, for any value of p. Mark in the point n is 1 and A equal to $1/(1 + p)$, which will fall on the sloping line. The result will look something like Figure 15-1.

To the right of the sloping line and below the horizontal line is the area in which must lie the graph of A versus n for semiautomatics. The graph will start from the point marked, and will be a curve convex upward. It cannot be a straight line or a concave upward curve, as in either of these two cases it would cut through the line A equals $1/p$ and this would violate the principle already established. As the curve progresses to the right, it becomes asymptotic to the line A equal to $1/p$. The above is common sense. It is not capable, however, of yielding precisely the one curve required for each value of p.

Ashcroft's (1) solution to this problem takes the form

$$A = (nY_n)/(1 + pnY_n)$$

in which Y_n represents the average number of stops attended in a stopping spell. In the work history of the machine there are only two sorts of spells—running and stopping. In the first, all machines, without exception, are running. In the second, at least one is stopped. The average number of stops which have to be dealt with between the end of one running spell and the beginning of the next is Y_n.

Evaluation of Y_n is difficult when the stop times are equal. In spite of this, Ashcroft calculated values of Y_n and A_n for all n values from 1 to 30. When the stop times have a negative exponential distribution, the difficulties are reduced a little. For this case, values of A_n have been calculated by this author for values of n up to 350. A part of this work is contained in Table 15-5. The graph in Figure 15-2 has been plotted for p equals 0.075, by using A values for the exponential distribution taken from Table 15-5. It will be seen that the curve fits into the frame formed by the lines A equals $1/p$, and A equals $n/(1 + p)$, so that these values pass the common sense test prescribed above.

The curve has been drawn asymptotic to the line A equal to $1/p$. This is not a mere assumption. That the curve must follow this path is a necessary corollary of the formula.

If

$$A = (nY_n)/(1 + pnY_n)$$

then,

$$pA = (pnY_n)/(1 + pnY_n)$$

The right-hand side is of the form $X/(1 + X)$, which can never equal 1 no matter how large X becomes. It will get nearer and nearer to 1 asymptotically, which is equivalent to saying that A will get nearer and nearer to $1/p$ in the same fashion.

In an actual case, of course, as n grows bigger p will necessarily increase too, since the walk-component is increasing. This point has not been overlooked, but it does not affect the argument. Ashcroft's formula passes every test that common sense can devise.

TABLE 15-4
Semiautomatics—Random Servicing of Random Stops. Stops and Servicing Data (as in Table 15-13 for cyclic servicing)

1. n	10	11	12	13	14	15
2. Walk per stop, minutes	0.060	0.066	0.072	0.078	0.084	0.090
3. Line (2) + t	0.360	0.366	0.372	0.378	0.384	0.390
4. Line (3) × q	3.60	3.66	3.72	3.78	3.84	3.90
5. p = line (4)/60k	0.071	0.072	0.073	0.074	0.075	0.077
6. A from Table 15-5	8.730	9.439	10.088	10.667	11.164	11.488
7. Overhead per hour	2.0	2.2	2.4	2.6	2.8	3.0
8. Cost per hour	3.0	3.2	3.4	3.6	3.8	4.0
9. Unit cost = lines (8)/(6)	0.343	0.339	0.337	0.337	0.340	0.348

In the preceding section productivity figures were calculated for certain time study data, and the optimum assignment was found, all on the basis of cyclic servicing. For the same time

TABLE 15-5
Productivity Factor, A, for Various Values of Cycle Ratio, p, and Machines Assigned, n

	Values of p						
n	0.073	0.074	0.075	0.076	0.077	0.078	0.079
1.	0.932	0.931	0.930	0.929	0.929	0.928	0.927
2.	1.855	1.853	1.851	1.850	1.848	1.846	1.844
3.	2.769	2.765	2.762	2.759	2.756	2.752	2.749
4.	3.670	3.665	3.660	3.655	3.650	3.645	3.640
5.	4.558	4.551	4.543	4.536	4.529	4.522	4.515
6.	5.428	5.419	5.409	5.399	5.390	5.380	5.370
7.	6.280	6.267	6.254	6.241	6.228	6.215	6.201
8.	7.107	7.091	7.074	7.057	7.039	7.022	7.005
9.	7.907	7.886	7.864	7.842	7.820	7.797	7.775
10.	8.675	8.647	8.619	8.591	8.562	8.534	8.505
11.	9.404	9.369	9.333	9.298	9.262	9.225	9.189
12.	10.088	10.044	10.000	9.955	9.910	9.865	9.819
13.	10.721	10.667	10.612	10.557	10.502	10.446	10.389
14.	11.296	11.230	11.164	11.097	11.029	10.961	10.893
15.	11.807	11.728	11.648	11.568	11.488	11.407	11.325
16.	12.251	12.157	12.063	11.969	11.874	11.780	11.685
17.	12.623	12.515	12.406	12.298	12.189	12.081	11.973
18.	12.926	12.803	12.680	12.558	12.436	12.314	12.194
19.	13.164	13.026	12.890	12.755	12.621	12.488	12.356
20.	13.342	13.192	13.044	12.898	12.754	12.611	12.470
$1/p$	13.698	13.513	13.333	13.158	12.987	12.820	12.658

study data, Table 15-4 shows the A values when servicing is random, and gives the cost results which point to the best assignment. Walk time per stop is assumed to be 0.006 times n minutes, and this time is added to the operator's work time before assessing p. When p is found, A_n is read from Table 15-5.

The answer again appears to be that an n of 12 will give the lowest unit cost, just as in the case of cyclic servicing. In these particular circumstances, cyclic servicing would be slightly lower in cost as compared with random servicing. The reason is that E in this case is 0.84 (equal to 10.088/12) as compared with 0.85 in the cyclic case.

The breakdown of one machine hour is:

Machine running	0.84
Machine at rest	0.16
Total	1.00

The operator is busy for 0.735 hour per hour (from $p \times A$) and idle the remainder, or 0.265 hour. B equals 0.735, and B/n is the average time spent on each machine; 0.735/12 equals 0.061. Thus, the machine rest item (0.16 hour per hour) splits into

Marow	0.061
Interference (by difference)	0.099

Interference

The foregoing is little more than an introduction to the general treatment of the subject. The basic problem of machine interference was solved by Ashcroft in 1948, after the present writer brought it to his attention (1). The problem occurs in numerous situations, and mostly these situations require the interpretation of data before the solution can be used. This is sometimes difficult. It is not possible to deal here with the interpretation of data and the application of the formula. Also, the tables which have been mentioned are somewhat bulky and have been published separately (7).

Ashcroft was not alone in tackling the mathematics of queuing, though he did it without knowing of any previous attempt. In the bibliography (11) compiled and published in April 1957 by Mrs. Vera Riley of the Operations Research Office of Johns Hopkins University are the names of almost 200 authors who wrote about 270 papers on the subject, and many of these were written long before 1948. Not all of them refer to machine interference, and only a few have outstanding merit, but their mere existence is proof of the great interest aroused by problems of this nature. The list was not complete when published, and there are now many more works waiting to be recorded. Of the papers dealing with the fundamentals of the problem of machine interference the following deserve notice either because of their value or the amount of publicity they have received.

The Wright formula (12) of 1934, based on work by Fry (3), was acknowledged by the author to be imperfect, but was put forward as a useful stopgap. It overestimates interference at all assignments (even indicates the occurrence of interference when an operator has only one machine). Fry's formula, on which Wright's is based, was devised for telephone circuits. It was built on the assumption that telephone calls from the various subscribers originated individually and collectively at random. At first sight this might appear to be analogous with the machine problem, with the machine stoppages represented by telephone calls; but it is plain that the analogy is false. The assumption would be true, for example, of six telephone booths in a public place, where a caller is just as likely to appear during an instant when five booths are occupied as when all are free. Under these conditions, (five booths occupied) the likelihood of a call being made from the sixth is six times as great as it would be if the others were free. If an operator has six machines, however, the likelihood of any one of them stopping is entirely independent of the condition of the other five. There is no true analogy and that is why the formula gives incorrect answers when applied to the machine problem.

Many writers have made use of the binomial theorem, but this leads to a solution which underestimates interference. This means, of course, that productivity is overestimated, and in fact when using this formula the curve of A versus n, if uncorrected, shoots through the upper limit $(1/p)$ of Figure 5-2, clearly indicating a basic error. Referring once again to the case of the operator with six machines, the stops and starts of these machines are not independent of each other.

However, the theory on which binomial distributions are constructed demands independence, such as would be obtained if each machine had its own operator. Therefore, the theorem is not truly applicable and hence the wrong answers. It appears there were at least three correct solutions before 1948, but none as simple or as complete as Ashcroft's. Chronologically they were the solutions of Khintchine (4) (1933), Kronig and Mondria (5) (1943), and Palm (8) (1947).

Of these, only Palm's solution is well known and has been translated (9). It deals mainly with the exponential case, but it has a section dealing with the effect on machine efficiency when two or more operators pool their efforts to reduce interference. Malcolm *et al.*, published a paper in 1955 (6), and followed it up by calculating some tables for the teamwork case.

In 1951, a thoughtful paper was published by Benson and Cox (2), dealing mainly with exponential distribution, but also introducing several exercises in interpretation.

To return to the industrial scene, it is agreed that machine interference is an evil, but it is an inevitable concomitant of cost cutting measures which place an operator in charge of several machines instead of just one. Once the multimachine assignment is made, interference occurs. To keep it in bounds, motion study followed by job training appear to be the best answer. Teamwork may be the next best, but teamwork calls for qualities of temperament which are rare, and the results are often disappointing. Very often, auxiliary workers are provided to relieve the principal operator. The cost of the auxiliaries added to the fact that there are two sets of interference, makes this a questionable procedure, but it may pay in some cases. Tables covering teamwork application have been published by Peck and Hazelwood (10).

Concluding Note

The question may be asked: "Why use mathematics? Why not measure interference as it actually occurs on the job?" A boy came to the bank of a swift stream and saw a trout. He threw in a cork to mark the spot and went to get his rod. When he got back, the trout had gone, and so had the cork.

The idea of measuring interference looks about as useful as the idea of "marking the spot" in a stream. It would be wearisome to list the chance causes that affect interference. Even with the most ingenious system of checks and corrections, there would be little hope of obtaining meaningful results. Calculated machine interference thus provides a more valid and logical basis than mere observation in providing proper time study allowance and determining expected efficiencies in multimachine assignments.

Where it is felt that none of the existing machine interference formulas would be applicable, a still further recourse to computer simulation has been suggested. By feeding a set of random numbers into an electronic computer, such as series of random deviates of exponential distributions, and permitting a random operation to "stop" various machines, the resultant effect on operator time loss and machine efficiencies might be evaluated from repeated "dry runs" on the computer. Such a procedure would, however, be more complex than the use of tables, as provided by O'Connor (7).

The purpose of this chapter is to discuss the various aspects of machine interference, so that the reader can use the method and approach to his own problems which he would consider most appropriate.

References

1. Ashcroft, H., *J. Royal Statistical Soc., Ser. B,* **12,** No. 1, 145–51 (1950).
2. Benson, F., and D. R. Cox, *J. Royal Statistical Soc., Ser. B,* **13,** No. 1, 65–82 (1951).
3. Fry, T. C., "The Theory of Probability as Applied to Problems of Congestion," *Probability and Its Engineering Uses,* Van Nostrand, New York, 1928, pp. 321–87.
4. Khintchine, A., *Matematiceski Sbornik,* **40,** No. 119, 73–84 (1933).

5. Kronig, R., and H. Mondria, *Physica*, **10,** No. 5, 331–6 (May 1943).
6. Malcolm, D. G., *J. Ind. Eng.*, **6**, No. 6. 19–26 (Nov.-Dec. 1955).
7. O'Connor, T. F., *Efficiency and Productivity Tables for Textile Machine Operation*, Institute of Textile Technology, Charlottesville, Virginia, 1959, p. 101.
8. Palm, D. C., *Industritidning. Norden*, **75,** 75–80, 90–4, 119–23 (1947).
9. Palm, D. C., *J. Ind. Eng.*, **9,** No. 1, 28–42 (Jan.-Feb. 1958).
10. Peck, L. G., and R. N. Hazelwood, *Finite Queuing Tables*, Wiley, New York, 1958.
11. Riley, V., *Bibliography on Queuing Theory*, Operations Research Office, Johns Hopkins University, Chevy Chase, Maryland, 1957, pp. 4, 7, 18, 24, 26, 28, 33, 44.
12. Wright, W. R., *Mech. Eng.*, **58,** No. 8, 510–14 (1936).

CHAPTER 16

Linear Programming*

Introduction

Linear programming was developed in 1951 by G. B. Dantzig (3), and represents a valuable method of analysis in the field of operations research. In particular, the technique permits the selection of an optimum combination of factors from a series of interrelated and interacting alternatives, each subject to limitations encountered in many industrial, economic, and military problems. In textile mills, linear programming has been found especially useful in studying the over-all operation of a mill or group of mills, so that those alternatives promising to lead to maximum over-all profit may be determined (2,5,6).

Production-Sales Coordination

The most useful application of linear programming in a textile mill is in the coordinating of sales goals with productive capacity, so that the combined efforts of selling and producing will lead to an optimized result in terms of final mill profit. Therefore, a mill producing a variety of styles would be a logical candidate for profit planning with the aid of linear programming. The reasons for this are as follows:

1. Each fabric style will cause different effects in the productive capacity of the various mill processing departments, as some

* The particular approach to linear programming described in this chapter, which has found acceptance in several mills, was first presented in a paper by N. L. Enrick, Textile Data Processing Association, Meeting at the Institute of Textile Technology, May 11–12, 1959.

styles will require more time for carding, roving, and spinning, whereas others will require more weaving and finishing time.

2. Each style has a different market potential and profit margin. Often, a style with good profit margins may be uneconomical to produce in large quantities, if its productive requirements in carding, drawing, roving, spinning, or weaving tend to create bottlenecks in one or several stages of processing. These bottlenecks would then disproportionately limit the production of other styles. Even though the low profit margin styles may be somewhat less profitable on a yard or loom-hour basis, they may, in terms of total yardage, contribute more to total profit if bottleneck limitations are absent. Linear programming evaluates the elements of production facilities, profit margins, and market potential with regard to each style. From this evaluation, a recommended optimum production combination that will yield maximum over-all profit, is obtained as the desirable goal for the coordinated efforts of sales and production.

In actual selling, the full attainment of the optimum combination of fabric styles and quantities may not be consummated, and in actual production the output may vary somewhat from the standard. However, knowledge of the optimum combination permits the sales efforts to be directed towards this goal. In addition, special promotions, price concessions, and inventory planning may be evaluated with a view towards the foreseeable maximum profit. Moreover, from experience accumulated with the use of linear programming, a practical percentage of actual versus ideal profit potential can be developed as a further planning aid.

Simplified Illustration

To demonstrate the principles of linear programming let us examine a simplified test case (2): Assume that a mill can produce two fabric styles, *A* and *B*. *A* yields a higher profit per yard, whereas *B* has a higher production rate per hour. Fabric *A* has a bottleneck in spinning and *B* has a bottleneck in weaving. The problem is: What quantity of each fabric style should be produced so as to yield maximum total profit for the combined output of both fabrics? Note that this problem cannot be solved just by producing only fabric *A*, since this fabric has a lower production rate and a bottleneck in spinning.

The solution, using simplified data for illustrative purposes, is given by the successive steps in Table 16-1, supplemented by the graphic determination of the optimum fabric combination in Figure 16-1. It is shown that for maximum profit 6.67 units of *A* and 6.67 units of *B* should be produced for the particular time period represented by the capacity data in *d* and *e*. The optimum profit of $233.3 results from $133.3 for fabric *A* and $100.0 for fabric *B*. In this particular example it happened that equal units of *A* and *B* must be produced for optimum profit; this is, however, merely a coincidence. Usually varying proportions of product are required.

TABLE 16-1
Example of Linear Programming Solution for Optimum Fabric Combination[a]

	Fabric A	Fabric B	Total
a. Profit per unit, dollars	20	15	
b. Spindle-hours required per unit	100	50	
c. Loom-hours required per unit	20	25	
d. Spindle-hour capacity			1000
e. Loom-hour capacity			300
f. Spun units per fabric at capacity (*d*/*b*)	10[b]	20[c]	
g. Woven units per fabric at capacity (*e*/*c*)	15[c]	12[b]	
h. Profit at spun capacity per fabric ($a \times f$)	200[d]	300[c]	200
i. Profit at woven capacity per fabric ($a \times g$)	300[c]	180[d]	180
j. Optimum units for both fabrics in combination (from Fig. 16-1)	6.67	6.67	
k. Optimum Profit at Optimum Fabric Combination, dollars ($a \times j$)	133.3	100.0	233.3

[a] For simplicity, only 2 fabrics are used. Actually, from 20 to 200 fabrics may be involved, and the optimum combination will then be found in a similar way.

[b] Bottleneck process for this fabric.

[c] Impossible because of bottleneck.

[d] Maximum profit if *only* this fabric is produced.

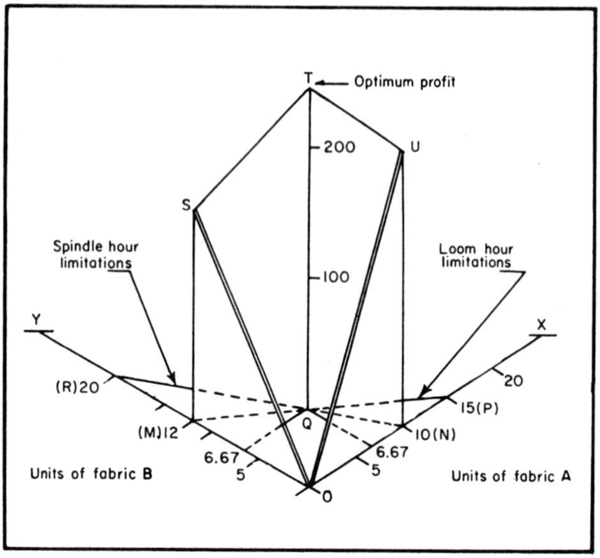

Figure 16-1. Fabric combination for optimum profit. The lines *RN* and *MP* cross at *Q*, corresponding to maximum profit. Perpendiculars from *Q* meet the base lines at the optimum quantities of 6.67 units for fabric A and 6.67 units for fabric B.

Practical Application

In practice, the mill will usually have more than two fabric styles, thus making it impossible to find a graphic optimum. The mathematical method of linear programming, however, permits handling of any practical number of mill styles that may be encountered. However, even with a dozen styles, the mathematics usually become so unwieldy as to call for the use of an electronic computer. Regardless of whether graphic, desk-computer, or electronic methods are used, the function of linear programming will remain unchanged: To discover that combination of fabric styles and quantities (often called "product mix"), which will maximize profit, after considering the effect of each style as regards profit per yard, production rate, and capacity utilization.

The illustrative example in Table 16-2 shows the optimum profit solution for a mill producing eight styles. First, there is

TABLE 16-2. Typical Linear Programming Problem and Solution Using Abbreviated, Simplified Actual Mill Data

	Marketable styles for season determined by sales department								Normal capacity per 120 hour week, in 1,000's of delivery hours[a]
	C	*D*	*E*	*F*	*G*	*H*	*I*	*J*	
Required data									
Dollar profit per 100,000 yards.	400	370	360	360	350	330	280	275	
Time needed to produce 100,000 yards fabric, in 1,000's of delivery hours.									
Cards	1.8	2.6	2.4	2.1	2.9	2.1	2.9	2.9	26.0
Roving	36.0	28.8	28.8	28.8	31.2	36.0	21.6	36.0	242.0
Spinning	720.0	648.0	720.0	648.0	648.0	720.0	792.0	720.0	7,200.0
Weaving	30.6	30.6	32.6	30.6	26.5	30.6	22.4	32.8	250.0
Maximum profit solution									
Optimum fabric combination for maximum profit, in 100,000 yards per week.	0.56	3.79	0	0	0	0	5.22	0	
Total profit per week, dollars.	224	1,402	0	0	0	0	1,462[b]	0	3,088[c]

Source: N. L. Enrick, *Computer Applications to Business and Business-Scientific Problems*, paper presented before Textile Data Processing Association, May 11–12, 1959.

[a] The optimum fabric combination would have remained unchanged, if in place of the normal 120 hours per week a lesser or higher number of hours had been used. However, the profit line would be changed proportionately, after allowing for overtime and other differentials.

[b] Note that this style, despite its low profit per yard, is actually the greatest contributor to total profit.

[c] Total profit, dollars per week.

a listing of the styles considered marketable and readily salable for the coming season, and within the normal productive capacities of the mill. For each style, the profit margin is provided. Usually, this profit is the operating or variable margin profit, representing the difference between selling price and variable costs. The cost department of the mill can readily provide such information. Table 16-2 also shows the productive capacity in all those processes that are potential bottlenecks, which in this example include carding, roving, and spinning in the spinning mill, and weaving in the subsequent processing. Other processes as picking, yarn preparation, slashing, and finishing were omitted, since capacity was considered more than ample.

The production rates are shown in terms of the number of thousands of delivery hours needed to produce 100,000 yards of fabric. The term "delivery hours" refers to spindle hours in roving and spinning and to productive hours per output position or delivery, in carding, drawing, and weaving. The output position in weaving is, of course, the loom. The data shown in the table are self-explanatory, except for the method of obtaining the optimum fabric combination. This combination is the result of linear programming calculations by using the method developed by Dantzig (3), as explained in a detailed and simplified manner by Noble (4) and Charnes, Cooper, and Henderson (1). The reader is referred to the last-mentioned book for a demonstration of the mathematical procedures.

Handling Market Restrictions

Often, it will be found that there are market restrictions on certain styles. This means that for some or all of the various types of fabric made by the mill, there is a limited market, so that prospective customers would not be expected to purchase more than a certain quantity of each style. This limit on quantity is known as a "market restriction." The illustration in Table 16-3 shows the basic problem data of Table 16-2, with an extra line for market restrictions as evaluated by the sales department. The resultant solution shows a somewhat lower attainable profit.

TABLE 16-3
Typical Linear Programming Problem and Solution Using Abbreviated, Simplified Actual Mill Data with Market Restrictions

	Marketable styles for season determined by sales department								Normal capacity per 120 hour week, in 1,000's of delivery hours[a]
	C	*D*	*E*	*F*	*G*	*H*	*I*	*J*	
Dollar profit per 100,000 yd.	400	370	360	360	350	330	280	275	
Time needed to produce 100,000 yards fabric, in 1000's of delivery hours.									
Carding	1.8	2.6	2.4	2.1	2.9	2.1	2.9	2.9	26
Roving	36.0	28.8	28.8	28.8	31.2	36.0	21.6	36.0	242
Spinning	720.0	648.0	720.0	648.0	648.0	720.0	792.0	720.0	7,200
Weaving	30.6	30.6	32.6	30.6	26.5	30.6	22.4	32.8	250
Market restriction, in 100,000 yards per week.	0.5	3.0					4.0		
Optimum fabrics for maximum profit, in 100,000 yards per week.	0	3.0	0	1.6	0.74	0	4.0	0	
Total profit per week, dollars.	0	1,110	0	576	259	0	1120[b]	0	3,065[c]

Source: An expansion of Table 16-2.

[a] The optimum fabric combination would have remained unchanged, if in place of the normal 120 hours per week a lesser or higher number of hours had been used. However, the profit line would be changed proportionately, after allowing for overtime and other differentials.

[b] Note that this style, despite its low profit per yard, is actually the greatest contributor to total profit.

[c] Total Profit, dollars per week.

This, of course, is as one would expect—the limitations on marketability have reduced the attainable profit.

Further Applications

It is a relatively simple matter to expand the application just shown to accommodate several hundred styles, so long as no problems with regard to computer storage are encountered in handling the mathematics. Moreover, many further features of practical production and sales can be handled readily by an expansion of the principles just demonstrated. These additional features include the following:

1. Market restrictions varying with price, so that lower prices would permit larger quantities of a fabric to be absorbed by the market.
2. Production on various types of equipment in a processing department, such as different types of looms running at different speeds. (In several actual mill applications, it was found profitable to run certain styles on looms that were wider than required and that produced at slower rates than the narrower looms.)
3. Use of purchased yarns in combination with mill-produced yarns.

Conclusion

Although it is a relatively new development, linear programming for profit planning is today a routinely used planning tool in a growing number of textile mills. Its application calls for close cooperation between the sales, production, cost, and industrial engineering departments. Thus, when linear programming points out bottlenecks in production, technological methods and standards changes and revisions must be considered, so as to overcome these bottlenecks at the lowest cost relative to market and profit potentials. When used in this manner, linear programming can be of great service to the mill.

References

1. Charnes, A., W. W. Cooper, and A. Henderson, *Introduction to Linear Programming*, Wiley, New York, 1953.
2. Eisemann, K., and W. M. Young, *Study of a Textile Mill with the Aid of Linear Programming*, American Society for Quality Control, Middle Atlantic Conference Transactions, 1958, pp. 9–20.
3. Dantzig, G. B., *Activity Analysis of Production and Allocation*, T. C. Koopmans, ed., Cowles Commission for Research in Economics, Monograph No. 13, Wiley, New York, N. Y. (1951), Ch. 21.
4. Noble, C. E., *Industrial Quality Control*, *13*, No. 9, 5–9 (March 1957).
5. Osawa, Y., *Industrial Administration Research Memorandum*, *1*, No. 5, 1–9 (June 20, 1955).
6. Enrick, N. L. Management Planning, New York, McGraw-Hill, 1976.

CHAPTER 17

Linear Programming—Further Applications

Introduction

The method of linear programming for profit planning given in Chapter 16, represents the most widely applied use thus far. However, many additional applications of linear programming can be made profitably in the textile industry. For example, a mill group having various producing, storage, and distribution centers, can use linear programming to minimize transportation and storage costs. Furthermore, many machine assignment problems can be solved for lowest cost by means of linear programming. A case history from the manufacturer of acetate yarn will illustrate the machine assignment problem.*

Machine Assignment Application

The linear programming application under consideration was concerned with the assignment of 10 acetate yarns among three types of dope metering pumps, so as to minimize repair and replacement costs. The production operation involved a multiplicity of extrusion machines, with each machine having 100 cabinets or positions, equipped with 100 metering pumps, all geared to a common shaft. By changing the pump revolutions and making other adjustments, different items may be produced. In this particular plant, 10 items or products were produced.

Three different types of pumps were in use, each exhibiting varying failure rates with the product being extruded. The costs

* This case history has been adapted from a paper by J. W. Cowdery, presented at Virginia Academy of Science Meeting, May 9, 1958.

TABLE 17-1
Linear Programming to Optimize Dope Metering Pump Assignment in Acetate Yarn Manufacture

	Products										Number of pumps available
	D	*E*	*F*	*G*	*H*	*I*	*J*	*K*	*L*	*M*	
Failure-repair costs per 100 hours, dollars											
Pump type A	3.36	0.84	2.87	6.72	11.34	1.68	0.91	1.47	1.12	4.76	293
Pump type B					5.81	0.21		0.84	0.70	3.08	97
Pump type C	0.70	2.10	0.70	0.77	6.37	0.77	0.91	0.98		4.34	194
Production required, pump-weeks	153	33	24	14	98	63	3	36	50	110	
Minimum cost assignment											Totals
Pump type A	0	33	0	0	0	63	3	36	50	108	293
Pump type B	0	0	0	0	95	0	0	0	0	2	97
Pump type C	153	0	24	14	3	0	0	0	0	0	194
Total cost, dollars (repair costs × minimum cost assignment)	107	28	17	11	571	106	3	53	56	520	1472

of repairing failures, usually by replacing pumps, corresponding to the various products, are given in Table 17-1.

By means of linear programming, it is now possible to assign pumps to each product in such a way as to minimize the total cost of failures. The solution, as shown in Table 17-1, yields a total cost of $1472 as the absolute minimum attainable under the given conditions.

The method of solution as given in many texts on operations research, (1) involves only simple arithmetic, but is too lengthy to be included here. It should be mentioned that in the actual application of the optimum solution a number of compromises with practical operating problems were necessary. However, by knowing the optimum assignment, it was possible to adjust actual assignment procedures to favor the optimum goal, which resulted in substantial cost savings.

Further Applications

It is apparent that the method of linear programming just demonstrated for pump assignments could also be adapted to other problems, as the optimum assignment of looms with various stop-rates to various cloth styles.

It is thus apparent that linear programming is a versatile and useful tool for the industrial engineer in his application of operations research techniques towards the minimization of costs and the maximization of production and profit. Although a growing number of linear programming applications are being made in textile mills, it is likely that a large number of potentially quite useful applications of this tool have not, as yet, been explored.

CHAPTER 18

Evolutionary Operation

Introduction

Evolutionary operation is a recently developed technique for the step-by-step improvement of a process, until it reaches an optimum level in terms of quality, costs, and yield. Developed by G. E. P. Box of Imperial Chemical Industries Limited (1), it is finding increasing application, including uses in the production of synthetic fibers. The method is applied to daily production by introducing a series of systematic small changes in the levels at which process variables are held. After each set of changes, the results are reviewed and new changes made, seeking through this evolutionary operation and development to gradually "nudge" the process to optimum levels.

The manner in which evolutionary operation or "Evop" can be used in the textile spinning and weaving processes was first described by Enrick.*

Example of Evop

The illustrative example used is taken from the sizing of plied filament yarns on the slasher. The primary purpose of the sizing operation is to condition the yarns for weaving, in order to minimize yarn breakage and loom stops and thus maximize weaving efficiency.

Existing mill practice involved settings yielding 6 per cent moisture content and 2 per cent elongation of the yarn, which is

* The material in this chapter has been adapted from N. L. Enrick, *Quality Control and Reliability*, 7th ed., Industrial Press, New York, 1977. (Reproduced by permission.)

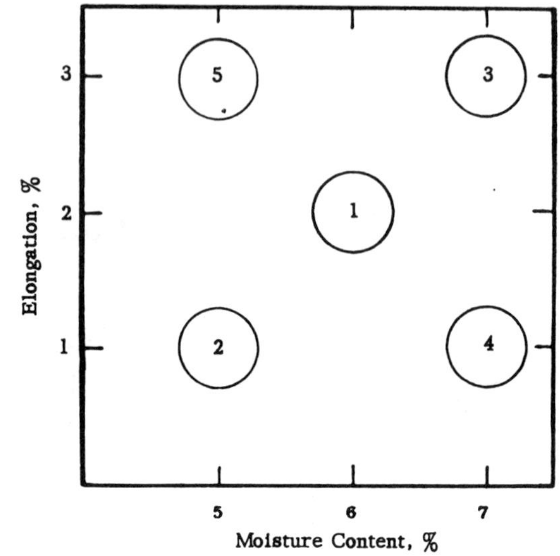

Figure 18-1. Design of investigation, comparing existing processing setup (1) against setups (2) to (5). Setups are represented by circles.

shown as setup (1) in Figure 18-1. It was decided to check this condition against the additional setups indicated on the diagram. These further conditions represent small changes, which were considered adequate to produce discernible effects on weaving efficiency, yet small enough to assure that no undue production or quality problems would arise.

Each set of operating conditions was run four times, resulting in four cycles, with weaving efficiencies as shown in Table 18-1.

TABLE 18-1
Weaving Efficiencies, %

	Setups				
	(1)	(2)	(3)	(4)	(5)
Cycle I	91	93	94	98	92
II	94	95	95	97	93
III	94	92	95	96	92
IV	93	96	96	93	91
Average	93	94	95	96	92
Range	3	4	2	5	2

The average of the ranges is 3.2. Since each of the setup averages is based on four determinations, we enter at this level Table 18-3, to find the control limit factor of 0.49. Therefore, ±0.49 times 3.2 or ±1.6 represents the extent of two standard errors around the setup average, in per cent weaving efficiency. We may call this ±1.6 more simply the "error limit."

The relative effects of moisture and elongation are estimated from the arrangement below, in which the setup averages are

TABLE 18-2
Relative Effects of Moisture and Elongation

Elongation, %	Moisture content, % (5)	(6)	(7)	Average	Effect of elongation
3	92		95	93.5	
2		93			−1.5
1	94		96	95.0	
Average	93.0		95.5	94.25	
Effect of moisture		+ 2.5			

TABLE 18-3
Factors for Determination of Error Limit

Number of cycles	Factor for error limit
2	1.25
3	0.68
4	0.49
5	0.37
6	0.32
7	0.28
8	0.25
9	0.23
10	0.20

Example: Assume 4 cycles were run, to yield an average range of 3.2. Then, the error limit factor corresponding to 4 cycles is found to be 0.49. Multiplying 0.49 times 3.2 yields 1.6. Therefore, the error limit is ±1.6.

Source: The table was computed for two standard errors of the difference between two sample averages, σ_d equals $\sigma(2/n)^{1/2}$ for comparison of means representing k equals 2 subaverages, such that the error limit is equal to $\pm 2\sigma/\sqrt{n}$. To permit use of the average range in place of the standard deviation, σ, allowance based on widely published conversion factors was made.

recorded to correspond to the relative positions of the setups in Table 18-2.

The effect of elongation, −1.5 in per cent weaving efficiency, is found by subtracting the elongation average of 95 from the elongation average 93.5. By similar subtraction of 93.0 from 95.5 for the moisture averages, the effect of +2.5 of moisture content is found.

The observed effect of moisture, +2.5, is well in excess of the error limit, ±1.6, and indicates strongly that increased moisture content of the yarn is beneficial for weaving efficiency. The observed effect of elongation, −1.5, is slightly below the error limit, but still large enough to suggest that reduced elongation is beneficial. In general, an observed effect which is below the error limit but still exceeds half its magnitude, may still be considered suggestive of a true effect. However, the smaller the observed effect in relation to the error limit, the greater is the probability that it may represent merely chance fluctuations of sampling and testing.

It is also interesting to compare the efficiency of setup (1), 93 per cent, against the average of 94.25 per cent for setups (2) to (5), which shows that the evolutionary operation just completed in its first phase has benefited and not harmed efficiency. A further evaluation may be obtained of the effect of interaction of moisture and elongation, by subtracting the average of setups (4) and (5) from the average of setups (2) and (3). In particular, 94.5 minus 94.0 yields 0.5, which is a negligible effect by comparison against the error limit of ±1.6.

We are now ready to decide on the next phase of evolutionary operation. Based on the observed effects, a good decision for further evolutionary changes in processing would be as follows: Run a further phase of four cycles. Select the center point, representing the new setup (1) at a somewhat higher moisture content of 6.5 and slightly reduced elongation of 1.5 per cent. Set up further processing conditions (2) to (5) at spacings of 1 per cent elongation and 1 per cent moisture in a symmetrical pattern parallel to that of Figure 18-1, but centered around the new setup (1). Then, after this new phase of operations has been completed, the results can be reviewed again and further phases planned accordingly.

Once experience has been gained with this type of operation, the program may be expanded. Additional criteria of processing cost and product quality may be included. Further processing variables, such as roll pressures, speeds and drying rates, may be incorporated in the investigations.

Organization for the Application of Evop

The application of evolutionary operation in a mill would require the cooperative effort of many parts of the production and staff organization. Initially, the industrial engineering department may arrange for the introduction of this technique by making suitable presentations to the production supervisory personnel concerned. Since factors of production, cost, and quality are involved in the determination of optimum levels, it may be practical to establish an evolutionary operation committee. This would consist of representatives of mill management, production supervision, industrial engineering, quality control, and cost accounting. This committee would act in an advisory capacity, aiding the technical superintendent and the other personnel concerned in the evaluation of the results of Evop and planning further investigations.

Many Evop programs will involve plant-wide investigations. For example, the effect of changes in drawing might be observed simultaneously, as regards yarn strength, spinning ends down, and loom stops. In order to minimize testing and observation time, the regular control records of the quality control laboratory must be used. Therefore, it is essential that quality control schedules of testing and inspection are maintained adequately, and that no other changes occur in the mill which would unduly interfere with the Evop program. Thus, the need for good organization and close coordination requisite for the successful conduct of evolutionary operation becomes apparent—as is true of all other industrial engineering and improvement programs.

Reference

1. Box, G. E. P., *Appl. Statistics*, **6,** 81 (June 1957).

CHAPTER 19

Inventory Control

Introduction

The textile industry is among those industries plagued, through all segments of manufacturing, by general conditions of overcapacity. The problems of overcapacity take on different aspects, depending on whether the market in which the mill's products are sold requires the manufacturer to build up inventories in anticipation of sales contracts, or whether sales contracts for future delivery can be negotiated before inventory is actually accumulated.

Fancy goods mills frequently enter each selling season with much, if not most, of their product line completely restyled. A highly seasonal business requires the accumulation, in the form of inventory, of a substantial part of their production, prior to sale in a market in which they must compete with dozens of other mills for their share of a widely fluctuating total industry volume.

For the manufacturer of simple cotton goods, it is true that styles change less rapidly and sales are less subject to seasonal fluctuations; however, it is a bargain-hunter's market in which each sale is separately negotiated. The question then becomes: How much of the mill's capacity—or its ability to deliver at a specified future time—should be committed today, and at what price? The over anxious mill manager commits too much capacity too soon and then finds himself turning down more profitable business because of inability to meet delivery requirements; the overly reluctant one finds himself with idle looms or spindles. In the hope of a better future he may use his excess capacity to produce inventory. But the ups and downs of the textile cycle often work the wrong way, and the accumulation of inventories only tends to weaken prices for the individual mill as well as for the industry as a whole.

The basic rule that management should follow is that the extra profits to be anticipated from producing inventory minus the cost of carrying the inventory must be on the positive side for each unit of added inventory.

Despite any notions to the contrary, carrying inventory is expensive if for no other reason than that money tied up in inventory reduces any mill's return on its investment. Consideration of alternative uses for money which otherwise would be locked in inventory is a basic factor in profitable operations in any industry, and the textile business is no exception.

Two typical applications of inventory control are presented.*

FIRST EXAMPLE—CONTROL OF INVENTORY PRODUCED IN THE ANTICIPATION OF SALES

General

In the case of fancy goods mills producing highly styled and relatively high priced fabrics, generally characterized by a strongly seasonal pattern of sales, each selling season represents in effect a separate venture. A new line of merchandise is prepared with very little carry-over of styles, patterns, or colors from one season to the next. Merchandise left unsold at the end of the season must generally be marked down substantially for disposal—often to a price which does not even cover variable costs.

This first system provides methods for measuring the risk of adding various quantities of different items to inventory. It consequently gives management a tool for systematically controlling the extent to which goods are manufactured in anticipation of customer orders, so as to achieve a minimum of end-

* Adapted with permission from a paper by D. B. Hertz and K. H. Shaffir, *Operations Research in the Textile Industry*, Arthur Andersen Chronicle, December 1959.

of-season write-downs, while providing adequate inventories to meet customer demands.

Net Expected Profit

The risk associated with inventory is measured by its net expected profit. For example, suppose we are dealing with an item whose regular selling price is $1.00 per yard, with a unit profit of $0.40 per yard over variable cost for each yard sold. If the item is left unsold at the end of the season it will have to be written down to a selling price of $0.40 per yard, resulting in a unit loss of $0.20 per yard.

Then, let us say that one specific yard of material, which we want to put into inventory, has a probability of 80 per cent of being sold during the regular season. The probability of its not being sold or being left over at the end of the season would, of course, be 100 per cent less 80 per cent, or 20 per cent. We therefore expect to make a unit profit of $0.40 per yard with a probability of 80 per cent, or a probable profit of $0.32 per yard. At the same time we expect to take a $0.20 unit loss with 20 per cent probability, or $0.04 per yard. The net expected profit of this particular yard of material is, therefore, the expected gain of $0.32 minus the expected loss of $0.04, or $0.28 per yard.

The selling price during the regular selling season, as well as the written-down selling price at the end of the season, are usually determinable from normal accounting records. The same is true of variable costs. It is clear, therefore, that the determination of net expected profit depends on being able to establish the probability that any particular yard of goods put into inventory today, will be sold before the end of the normal selling season.

Seasonal Order Build-Up

The first step in arriving at this probability is to consider the way in which sales for the style line, as a whole, develop during the course of the selling season. A line for this purpose

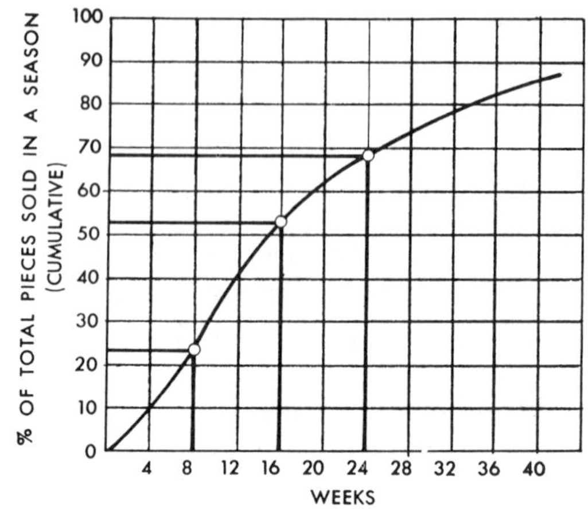

Figure 19-1. Typical sales pattern.

includes all items manufactured for sale in a single distribution channel at roughly the same price range. For example, men's flannel shirting selling in the range of $1.00 to $1.20 per yard would be considered as one line. An item in such a line would be a specific pattern and color combination such as red and gray plaid flannel shirting material.

One may express orders received to date at the end of each week in terms of the percentage of total sales for the line for the entire season, and then plot the percentage figures obtained against the corresponding week numbers as shown in Figure 19-1. (For purposes of simplicity, we will refer to orders received as "sales.") In general, one will find a sales pattern developing slowly at the beginning of the season, building up rapidly as the season gets under way, and then leveling off again towards the end of the season. For example, at the end of the 8th week of the season, one might have sold about 23 per cent of the total sales of the line for the entire season, at the end of the 16th week about 53 per cent, and at the end of the 24th week about 68 per cent. This basic pattern varies from line to line, depending primarily on the length of the selling season and also on the specific market served, such as men's wear, women's wear, piece goods, etc. Given the type of line, it is necessary then, to find the length of the selling season in order to determine the basic sales pattern.

In some markets the duration of the season is quite constant and predictable from year to year. In others it varies and must be reestimated each season. The procedure used for doing this involves a continuous comparison of orders received to date for the current season with various "standard" patterns representing different lengths of season, to determine which standard most nearly corresponds to the actual.

Probability of Sales

Individual items in the line will, of course, deviate from the basic pattern of the line as a whole, as shown by the black line in Figure 19-2. The basic sales pattern for the line gives a means of estimating the total sales for the entire season, on the average, based on the amount which has been sold to date. For example, we know that for this particular line we should expect to sell 40 per cent of the season's total by the end of the 12th week. Therefore, if we have sold 1,000 pieces to date, we may expect to sell a total of 1,000 divided by 40 per cent or 2,500 pieces for the line for the season as a whole. Another way of saying this is that half the items in the line have sold more than 40 per cent,

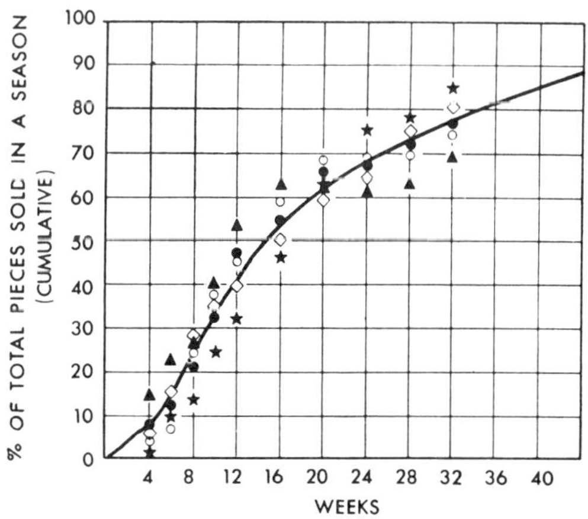

Figure 19-2. Deviation of individual items from sales pattern of a line (item A = ●, item B = ○, item C = ◇, item D = ★, and item E = ▲).

or 2,500 pieces for the line for the season as a whole. In terms of probability we can say that we are 50 per cent sure that sales to date for any one item represent not more than 40 per cent of total sales for the season. Conversely, we are also 50 per cent sure that sales to date for any one item represent not less than 40 per cent of total sales for the season.

The basic sales pattern can, therefore, be considered as a 50 per cent probability limit on sales of individual items in this line. Sales of an individual item, of course, cannot be predicted with as much accuracy as for the line as a whole. However, by analysis of past sales records, such as are shown here, the probability that any one item will deviate by more than a specified amount from the over-all pattern of the line can be determined. For example, it can be determined that by the end of the 16th week, 90 per cent of the items would have sold 60 per cent or less of the seasonal total. Conversely, 10 per cent of the items have sold more than 60 per cent of the seasonal total. We determine this 90 per cent limit by looking at the data for individual items such as are shown in Figure 19-2 and finding that 90 per cent of the items did, in fact, have sales of less than 60 per cent of the seasonal total by the end of the 16th week and 10 per cent had sales of more than 60 per cent of the seasonal total.

Therefore, we can say with an assurance of 90 per cent that total season sales will be at least 1,000 divided by 60 per cent, or 1,667 pieces. This, of course, is a more conservative estimate than the 2,500 pieces anticipated at a 50 per cent assurance level.

With a 70 per cent probability limit, the 1,000 pieces sold to date might represent 50 per cent of season total, giving an estimate of 1,000 divided by 50 per cent, or 2,000 pieces. This, of course, falls between the 90 per cent and 50 per cent estimates of 1,667 and 2,500, respectively.

Note that at the 100 per cent probability limit we can be 100 per cent sure that sales to date represent not more than 100 per cent of season total, or that total season sales will be at least 1,000 pieces divided by 100 per cent or 1,000 pieces.

For use in the inventory anticipation system, tables are prepared of incremental sales ratios which, when multiplied by season sales to date for a specific item in the line in a specified week, give the number of additional pieces expected to be sold with a specified average probability. Separate tables of this type are provided for different lengths of selling seasons for each of the lines. These estimates directly take into account current business conditions in the sense that the higher the sales generally, the higher the estimates, being based on current sales to date.

Presentation to Management

The information gathered must now be put together in a form in which management can use it. This is illustrated in Table 19-1. A separate table of this sort is prepared each week for each class of loom, listing all currently active styles which can be produced. Looms within a class can generally be used interchangeably on any of the styles listed for that class. For simplicity's sake, only two styles are shown here.

The lefthand portion shows the style number, the current profit contribution of the style in dollars per loom-week, and the orders received for the season to date, expressed in loom-weeks of production.

On the right there are, for each style, a series of forecasts corresponding to different levels of net expected profit (NEP) contributions. In this case, NEP levels are \$10, \$15, and \$25 per loom-weeks.

These levels are arrived at by first determining the probability of sale required to yield each level of NEP. Style X-1, for example, has a current profit of \$20 per loom-week. To achieve a net expected profit of \$15, therefore, requires a probability of sale of 75 per cent. For style X-14, only 58 per cent probability is required to yield \$15 NEP.

Given this percentage, a forecasting factor for the current week is selected from a series of tables, depending on the style's seasonal pattern and its variability. This factor, when multiplied by the orders received to date, give the forecast of additional sales for the season at the specified probability and NEP levels.

To arrive at the quantity actually available for scheduling of looms, the amount already on hand or in process must, of course, be deducted from the forecast.

TABLE 19-1
Quantities Available for Scheduling (Expressed in Loom-Weeks of Production)[a]

Item number	Current profit, dollars per loom-week	Orders received to date in loom-weeks		Net expected profit contribution, dollars per loom-week: $25	$20	$15	$10
X-1	20.00	2,410	Probability		100%	75%	50%
			Forecast		0	460	1,300
			On hand and in process		410	410	410
			Available		0	50	890
X-14	25.80	1,250	Probability	97%	78%	58%	39%
			Forecast	50	350	850	1,760
			On hand and in process	640	640	640	640
			Available	0	0	210	1,120
Total available				0	0	260	2,010

[a] In this example, the profit on after-season sales (for all styles) is taken as zero (break even on variable cost).

It is now possible to add up the total amount available for scheduling for all styles. At the $10 level there are 2010 loom-weeks, at the $15 only 260 loom-weeks, and at the $20 level none. The number of looms to be scheduled will determine the total number of loom-weeks required, and consequently, the level of expected profit at which to operate. The way in which loom capacity should be allocated to individual styles can then be read from the table. If 260 loom-weeks are to be scheduled, for instance, 50 would be allocated to style X-1 and 210 to style X-14.

Summary of Inventory Anticipation System

Thus, Table 19-1 gives management the basis for allocating currently available loom capacity to each individual style or item, taking the following factors into consideration for each item:

1. Current profitability.
2. Expected mark-down if left over at end of season.
3. Historical sales patterns for the style in prior years.
4. Sales pattern developing in the current season.
5. Amount of orders currently on hand or received to date.
6. Predictability of sales for the season, based on deviation of prior years' sales from what would have been forecast had this method been used.

The system is self-correcting in that a new forecast is made each week (see Figure 19-3). As sales orders are received at the mill, they go to a production control clerk who posts sales both for the line as a whole and for each individual item in the line. The data for individual items, in conjunction with the length of season as determined for the line, are used to make item sales forecasts at different levels of probability. These are then posted on a summary sheet from which production orders can be issued. Allowance must also be made, of course, for the posting of production orders so as to keep inventory records current.

Management is apprised at all times during the season of the odds of selling enough goods to fill the looms profitably, and of how to allocate looms among styles so as to get the best odds.

If the capacity of a loom class for the season is greater than the quantity available for scheduling at or above zero net expected profit (break even), management knows that continued operation at capacity with the present styles will result in unjustifiable accumulation of stock. It may then take steps to curtail production on specific styles or transfer styles from another loom class where possible, and in this way control inventories.

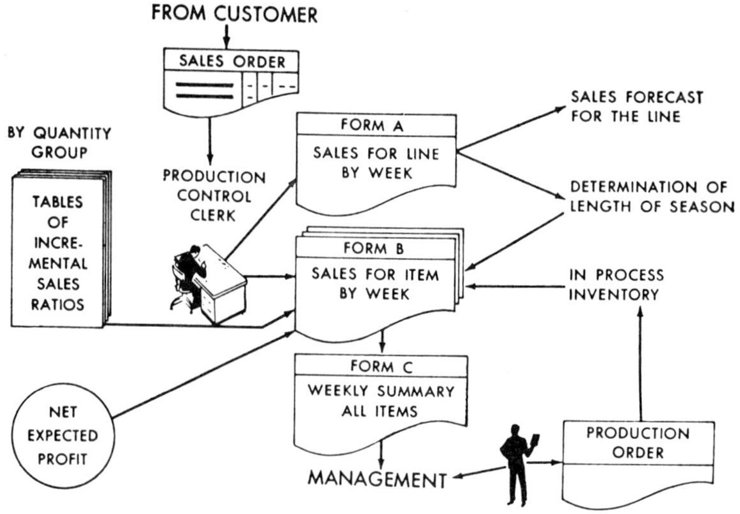

Figure 19-3. System flowsheet.

It should again be pointed out that this system is not designed to eliminate judgment. Rather, it is intended as a basis for the application of judgment by marketing and production personnel. The system does not eliminate write-downs or lost sales—it attempts to balance inventories so as to keep write-downs at a level which is economically justifiable. The system does improve the odds of having the right material on hand throughout the season.

SECOND EXAMPLE—MAKING MORE PROFITABLE SALES: A SYSTEM FOR SETTING "REFUSAL PROFIT" LEVELS

The second example deals with a method for controlling the price at which orders are accepted for delivery during any specified future month. This method is applicable in the case of most cotton fabrics and blends, or generally, whenever price is separately negotiated for each order.

The basic situation may be described as follows: The mill manager's objective is to sell the production of each future month at as high a profit as possible. The total quantity which can be delivered up to any specified date is, of course, determined by production capacity. However, the mill manager has, as a rule, many alternative ways of selling a given month's production. First, he can sell it in the form of many possible combinations of the several styles that his looms can produce. Second, he will be given many opportunities to sell—in the form of offers from his customers—starting perhaps a year in advance and continuing up to the end of the month in question.

To accept the first offer he gets is obviously psychologically bad and may result in selling too cheaply. On the other hand, if he refuses too much and too often, he will be left with unsold capacity, or in any event will often find that he turned away business he later wishes he had taken.

It is obviously impossible to do a perfect job. However, an estimate or "forecast" can be made each week of the amount of business—in terms of loom-weeks of production—which will be available for the mill manager to select.

Furthermore, it is possible to estimate the relative amounts which will be available at various levels of profitability (in terms of dollar profit per loom-week), and to compare these estimates with the amount of production capacity remaining to be sold for each month or quarter. On the presumption that the most profitable business estimated to be available will be accepted and the rest refused, the "refusal profit" can be determined as that level above which the forecast shows just enough available business to cover capacity. These estimates must, of course, be revised as new information becomes available, and the refusal profit level adjusted continually to keep the new orders accepted coming in at the proper rate.

Inventory Accumulation Versus Advance Sale

There is also the possibility, of course, of not selling some portion of the available productive capacity ahead of time, but rather committing it to inventory. This only means that a correspondingly greater amount will have to be sold in a subsequent quarter. In other words, accumulation of unsold inventory is equivalent to betting that business will be significantly better at a future time. However, an upturn in future business—in terms of either price or volume—is generally signaled by current offers to buy at a favorable price for distant future delivery. Such offers are in fact the best foundation for predicting an upturn. This means that in those cases where there is an argument for converting current excess capacity into unsold inventory, it is generally also possible and safer to cover this excess capacity through sales for more distant delivery. The system described here is designed to accomplish this, i.e., to keep the rate at which orders are accepted at a level which gives reasonable assurance that total loom capacity will be covered by sales orders prior to production, so that the only inventory required will be sold inventory. This does not preclude spot sales where the opportunity for extra profit presents itself. Such sales can be made by using inventory which, though covered by existing sales contracts, can be replaced by the time its delivery comes due.

Basic Concepts

To understand how the system actually works a basic concept needs explaining:

It is useful to group styles into categories according to their end use and their profitability expressed in dollars per loom-week. All sales orders for a category can then be expressed in

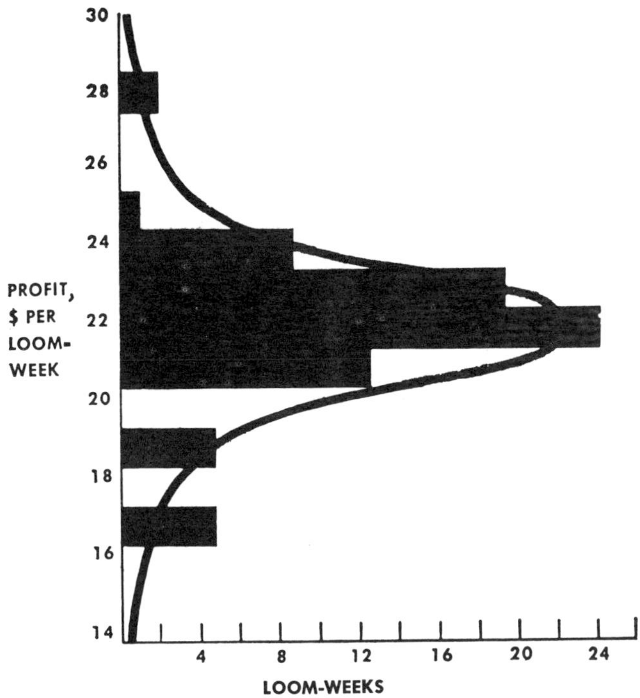

Figure 19-4. Representation of a group of sales orders in terms of a normal distribution.

2 Loom-weeks at	$28 per loom-week	=	$ 56
1	25	=	25
9	24	=	216
20	23	=	460
25	22	=	550
13	21	=	273
5	19	=	95
5	17	–	85
80 Loom-weeks			$1760

Average = $22 per loom-week; standard deviation = $2 per loom-week (99 per cent between $16 and $28 per loom-week).

terms of a normal distribution. This is illustrated in Figure 19-4. Eight sales orders, ranging in profitability from $17 to $28 per loom-week, and totaling 80 loom-weeks are represented by a "normal curve"—the bell-shaped curve well known to statisticians. We can say that these eight orders are reasonably described by (1) the total quantity of 80 loom-weeks; (2) the average profitability of $22 per loom-week; and (3) the standard deviation of profitability of $2 per loom-week. These facts indicate that based on the normal distribution curve, about 70 per cent of the total volume falls within $2 of the average profitability, and 99 per cent within $6 of the average.

These eight orders could represent eight different styles or eight different sales of the same style, made at different times. In either event, they can be described in terms of their total quantity and of the average and standard deviation (SD) of their distribution of profitability.

The reason for making forecasts in the form of normal distributions is that this makes it possible to easily estimate (by use of standard statistical tables) the fraction of the total amount of forecast business that can be expected above a given level of profit contribution. Suppose that Figure 19-4 did not represent sales orders already received, but rather a forecast of orders to be received the next quarter. The standard tables would indicate that 84.1 per cent of the total 80 loom-weeks, or 67.3 loom-weeks, would have a profitability above $20. The actual number in our example is 70 loom-weeks, which indicates a fair estimate.

If total capacity for the next quarter were, in fact, 67 loom-weeks, our forecast would show that a sufficient amount of business should be available above the $20 profit to fill that capacity, and that business yielding less than $20 should, at this time, be refused. In actual practice, the calculation is slightly more complex because several categories of styles—each having a different average profitability and standard deviation—must be considered in fixing the refusal profit level for each class of looms. The basic method is, however, the same.

Forecasting

The manner in which quantity and profitability of each category are forecast varies, of course, from mill to mill. In general, quantities are projected on the basis of historical sales build-up patterns applied to the amount of unfilled orders currently on hand, in a manner similar to that described for the inventory anticipation system.

Profitability of future business may be estimated from current or recent orders taken, or from published price indices. A wide variety of methods is currently in use.

Most important, a new set of forecasts is made each week, continually adjusting the refusal profit level on the basis of current experience. Thus, if more than the normal amount of business is currently sold, the system yields forecasts of higher quantities for the future, resulting in a higher refusal profit level so as to eliminate more of the low profit business. Conversely, if a time period goes by without sales, the amount of future business forecast is gradually reduced and the refusal profit drops, to encourage acceptance of more business. Changes in the profitability of current business similarly affect the refusal profit level.

Here again, various modifications in the system are required to adjust for such factors as imbalance between spinning, weaving, and finishing operations, in order to take into account the cost of commission finishing or purchasing yarn on the outside.

Refusal Price

For effective use in day-to-day price negotiations, refusal profit levels expressed as dollars per loom-week, must, of course, be translated into refusal prices for specific styles, taking into account costs, loom speeds, and sales commissions.

Thus, each style has a different refusal price in terms of cost per yard. In addition to this, the system provides an "acceptance price" based on the forecast average profit level of the loom class. In general—subject, as always, to the merchandiser's judgment—an offer to buy would be accepted above the acceptance price and rejected below the refusal price. Between these two price levels (which are sometimes fairly far apart) lies an area in which judgment, based on the merchandiser's knowledge of the customer and market, must provide the final answer.

Inventory Control Through Pricing

Separate refusal profit levels are established for delivery in each future quarter. If the current quarter's refusal profit levels are substantially lower than those of the next quarter, a procedure is provided for considering both quarters together so as to allow for the production of some of next quarter's deliveries in the current quarter. This procedure takes into consideration the cost of carrying inventory over for one quarter, and limits the extent to which this is done based on the economics involved.

At the same time, it provides for the acceptance of less profitable business for distant future delivery, to the extent that it is advantageous to do this and thereby avoid loom shut-downs.

CHAPTER 20

Computer Applications in the Industrial Engineering Function

Introduction

Although most industrial engineering functions carried on as a routine in the textile mill would not warrant the purchase of an electronic computer, once such a computer is installed there are many applications of industrial engineering calculations which can and should be "computerized." Machine calculations will not only reduce the work of desk calculations and the tedium and drudgery connected with this activity, but also provide the following cost saving advantages:

1. Time savings in obtaining results of analyses, summaries, and computations.
2. Virtual assurance that there will be no errors in the calculations, so long as the data input into the machine has been correct.
3. Reductions in manpower and costs.
4. Ability to place new projects, new analyses, and special summaries on machine computations, where previously the time and money involved in desk calculations made such applications impossible.

In the following, some key industrial engineering and operations research calculations suitable for computerization will be presented. The material is based upon survey data obtained by the Textile Data Processing Association, supplemented by experience and observations of the editor in his work with textile mills.

Incentives Application

The calculation of standards in connection with incentives application can often be very complex, corresponding to the need for close integration of operator-machine cycles in multi-machine assignments in practically all textile processing operations. Although some mills have considered these calculations too complex for computer applications, others have reasoned that it is precisely this complexity which would yield considerable cost savings through computerization. Thus, industrial engineering standards can be, and have been, placed on computer applications, as reported in the aforementioned survey by the Textile Data Processing Association.

Supplementing the standards calculations, the following historical and control data may be obtained, as concerns the actual operation of incentive pay: weekly and monthly payrolls, labor distribution, reconciliation of actual with standard labor costs, comparative machine and operator efficiencies, ends down and loom stop rates by styles, and costs per pound by styles.

Quality Control

Considerable advantages are derived for quality control work if basic data are fed into a computer, and summaries by various groupings are then obtained. For example, records of cloth grading and recheck grading may readily be analyzed in summary form according to the following categories:

1. Major and minor defects, by styles.
2. Major and minor defects, by styles and looms.
3. Comparison of defects by styles, looms, and weavers.
4. Comparison of grading report defects with recheck grading defects found.
5. Grader proficiencies, as a per cent of correctly graded defects versus recheck grading defects analyses.
6. Correlation of defects rates with raw materials used.
7. Comparison of customer complaints with grading report data.
8. Analysis of defects by types.

9. Analysis of defects by types, weavers, and loom fixers responsible.

Almost all types of quality records, including data from the control testing laboratory, can be subjected to further statistical analysis by means of repeated grouping and regrouping of the information on an electronic computer. Although this same information could also be obtained by manual methods, the time and cost involved are usually prohibitive.

Production and Inventory Control

Concurrent with the development of payroll data, various production reports can be obtained, showing actual pounds produced at each process, as well as waste, yield, machine efficiencies, productivity, shrinkages, elongation, and contractions. Data may be summarized by styles, departments, and often the operators involved.

Such data as warps on floors and warp yardage on the looms will assist in the planning of the flow of production, and in the maintenance of minimum in-process inventories consistent with adequate safety levels.

Inventories of supplies, maintenance stores, raw materials, goods-in-process, chemical additives, and finished goods, can be controlled readily at near optimum levels through use of computer accumulated information, and the application of operations research techniques to the data obtained. Usually, savings in inventory carrying costs and the reduction in obsolescence and deterioration of supplies and lowered accumulation of "distress goods" are among the greatest benefits to be obtained from computer applications. It is for this reason that inventory control has, in many instances, been the first factor to be placed on computers.

Production-Sales Coordination

Sales analyses by styles, customers, territories, salesmen, and seasons, essential to effective management and control of selling operations, may be obtained quickly from computer applications.

In multistyle mills sales forecasting may best be carried on with the aid of a computer, and the analysis of past performances correlated with future expectations. Linear programming for profit planning is next used, together with inventory analysis, in coordinating sales, production, and inventories. This permits weekly planning by management, with the objective of optimal integration of production, inventory accumulation, and sales efforts with current market potentials and productive facilities in the mill.

Practical Aspects of Electronic Data Processing of Industrial Engineering Applications

In preparing analyses and summaries of costs, production, productivity, and quality it is essential to confine all efforts to only those reports which will be helpful to mill management, the analytical staff, and mill supervisors in actually evaluating and controlling operations, from a viewpoint of achieving corrections and improvements in operations where required. Excessive data, redundancy of information, and figures of marginal interest will not only add to administrative costs, but also detract from the over-all evaluation and control program. Similarly, when incentives applications are aided through computer calculation of rates, care should be taken that all essential calculation steps are printed out by the machine, so that the procedures used can be readily checked by all persons concerned.

As the computer is usually under the administrative control of the accounting department, it is desirable that an experienced industrial engineer, who is mathematically inclined, be in charge of planning, programming, supervising, and coordinating all data processing concerned with the industrial engineering phases of analysis and control.

PART IV

MANAGEMENT INFORMATION SYSTEMS—INSTALLATION, OPERATION, AND CONTROL

CHAPTER 21

Management Information Systems Overview

Introduction

Property selected, installed and used, the computer in today's textile organization is a practical tool which saves routines of drudgery, provides relevant information quickly, sharpens the decision-processes from line-management on up to top management, and permits management and staff alike to devote more time and effort on creative and innovative projects. The ability of the organization to deal with change—technical, market-wise, and socioeconomic and political—determines its long-range viability and strength.

Purpose of Management Information Systems

In this Part IV on "Management Information Systems," concern is with the practical control of operations, using the computer and its far-flung terminals throughout the textile mill and the textile organizations, as a management information aid. "Well informed is well decided." To the degree that the computer-based information system provides relevant information, omitting needless detail, it will provide real value to the decision-makers. The following aspects of computer information systems[1] will be covered:

– The nature and role of information systems.

[1] Throughout, the terms "Computer Information Systems" (CIS) and "Management Information Systems" (MIS) will be used interchangeably with identical meaning. In effect, "Computer Information Systems" is the more specific terminology, but "Management Information Systems" has gained a wider acceptance.

- How to determine the feasibility of proposed systems as an aid in developing an effective final system.
- Means whereby costs can be weighed against expected benefits, thereby justifying installation and operation of the system.
- Achieving, throughout the organization, motivation to promote use of the computer and to realize the benefits so attained.
- Developing an overall view and appreciation of information systems, recognizing not only potential benefits but also the pitfalls to be guarded against.

Information Flows and Decision Processes

The general framework of information flows within which a management information system operates is illustrated in Fig. 21-1, on the executive decision process. The constant influx of data in all areas of interest—production, marketing, finance—is converted to the essential information content. Within the context of the decision-problems to be resolved, there is a weighing of alternatives (such as promotion approaches, pricing system, models and styles of fabrics to be offered, inventory policies to pursue) against limitations applicable (market requirements, cost factors, capacities of spinning, weaving and converting equipment, kinds of financing available, and other pertinent considerations). Qualitative and quantitative factors are considered before a decision on the most desirable course of action is reached. The effects of these decisions—on production and productivity, market-success, financial status, and other areas—are subject to review and analysis, thus forming the input to the next problem-solving and decision-making cycle.

In subsequent chapters, detail is presented on how to go about selecting, installing and operating a cost-effective system of computer-generated operating and control information, which will serve profitably in managerial and staff tasks and enhance the quality of the decision processes of the organization.

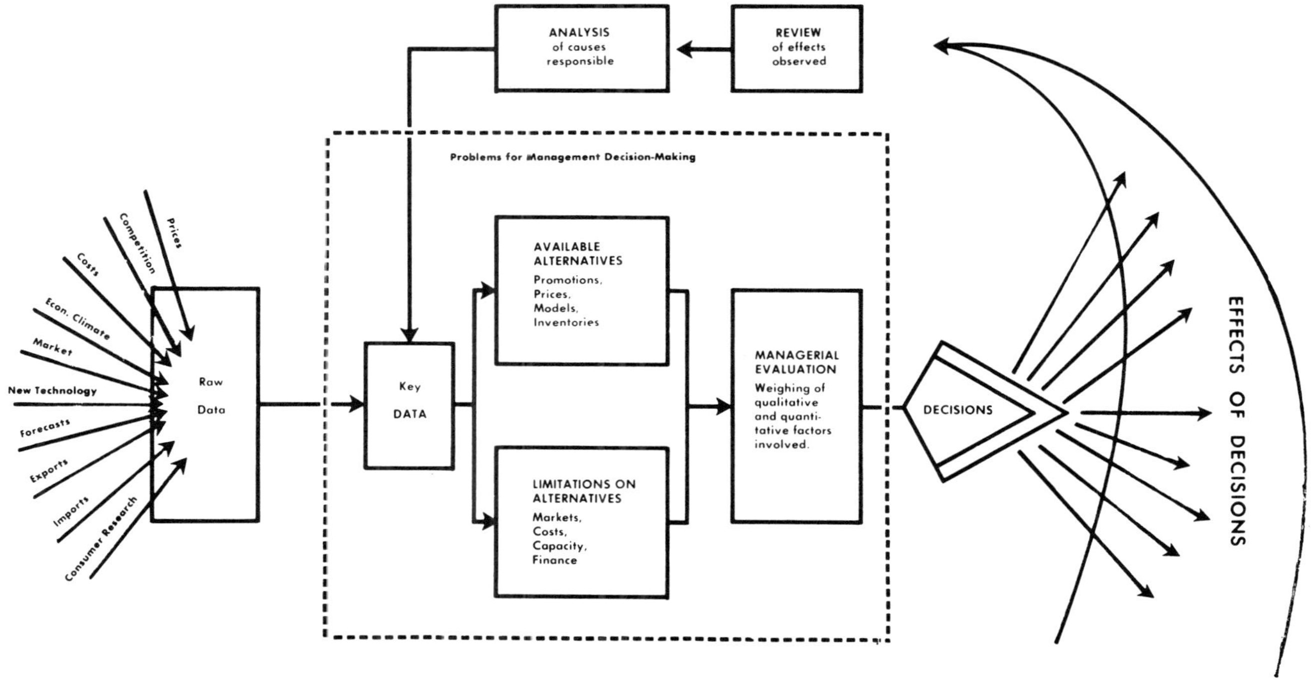

Figure 21-1. The decision process. Incoming data, converted to essentials, are evaluated against available alternatives and limitations. From an evaluation of these factors, decisions are formed. A feedback system provides for review and revision of decisions.

The Role of Computers

Computers and automatied data processing are becoming ever more deeply entrenched in modern life. Despite occasional failures and instances of back-tracking, the overall thrust is towards increasing reliance on computers in commerce, industry, government, the social and physical sciences and the fine arts.

A consumer who buys a textile product can be reasonably sure that a computer has aided in much of its design, that computer-automated controls have governed a major part of production, and that a system of electronic data processing has guided scheduling and the coordination of sales, inventories and production. The record keeping, accounting, billing and related work of practically all textile businesses is done through computer processing.

Both major and minor aspects of the everyday business of living increasingly are affected by computerization.

Business-Industrial Developments

Computer applications must be justified in terms of usefulness, profitability and time-saving features. Several factors have been instrumental in bringing about increasing reliance on automated data processing by small and large business enterprises alike:

1. The cost of using computer services or of owning one's own facility has been falling rapidly. Advances in technology, particularly miniaturization and automated production, have made this success possible.

2. We have learned how to develop general programs that are immediately applicable to a wide variety of business and government needs in such areas as payrolls, accounts receivable, accounts payable and inventory records. Reliable output is attainable, with benefits in cost and time savings at no sacrifice in reliability.

3. A good Management Information System (MIS) installation will provide decision makers with data that serve to vastly improve operations. Such questions as: "What is the

minimum inventory of various fabric styles needed, without sacrificing sales potentials?" or "How do I schedule spinning, weaving and finishing production to minimize processing delays and accurately meet customer demand dates?" are properly analyzed by a good MIS.

4. There is great flexibility of application. A company has many alternatives, ranging from simple batch to remote batch processing, time-sharing or owned facilities, and various combinations of these and others. Increased flexibility means increased applicability of automated data processing for all businesses.

Almost daily there are new improvement and refinement of methods and techniques, further cost reductions and time savings, and additional simplification of usage.

Alternatives

Four major choices of data processing are available, involving (1) plain batches, (2) remote batches, (3) time sharing and (4) on-site computing, as summarized schematically in Fig. 21-2. The following further detail is significant.

- A batch represents data of a similar nature, such as raw materials inventory, permitting processing as one group at one time with one computer program.
- Batches are processed by a service bureau. Plain batches are sent back and forth by mail, car or messenger. Remote batch processing relies on send-receiver terminals at user and processing site, with transmittal on telephone, teletype or other wire. While ordinary batch processing involves fees for work done only, remote methods incur fixed costs for the installation and use of terminals and wires.
- Time sharing involves direct connection, again by wire, between the user and the computer itself. However, while batches usually have to wait (hours, days or more, as arranged) until they can be handled, time-sharing computers handle inputs immediately. In fact, several users' data are processed simultaneously at any one time.

<table>
<tr><th colspan="3">COMPUTER USE ALTERNATIVES</th></tr>
<tr><th>User Method</th><th>Transfer</th><th>Outside Facility</th></tr>
<tr><td>Plain batch of data</td><td>car</td><td rowspan="2">Delayed processing
(Hours, days or week)</td></tr>
<tr><td>Remote batch</td><td>send-receive terminals</td></tr>
<tr><td>Time sharing</td><td>direct-to-computer line</td><td>Immediate processing. Computer services several terminals simultaneously</td></tr>
<tr><td>On-site computer</td><td>none</td><td>none</td></tr>
</table>

Figure 21-2. Computer use alternatives.

There is no waiting. The user's send-receive terminal works as though the computer were in an adjoining room responding directly to the user's inputs.

– On-site computers are expensive not only in terms of fixed costs, but also maintenance. From 0.5 to 2.0 percent of a company's gross sales revenue may be involved.

Combined Uses

A recent trend is to design automated data processing (ADP) systems that use a combination of methods, such as (a) plain and remote batch, (b) plain batch and time-sharing, (c) remote batch and time-sharing and (d) owned computer and service bureau batch processing. This saves money. For example, data compilations for which fast responses are required may go via remote batch; other items permitting a week or two of turn-around time go by car to the service facility. Or, if a company's own computer is overloaded, unable to handle certain types of specialized analyses, or inadequate in speed or memory capacity, the company can resort to a supercomputer at a service facility.

The starting point for a small or medium-sized mill would logically be batch processing at a service bureau installation.

There are many reasons for this: (a) costs can be predicted with relative accuracy and contrasted against values to be gained, (b) costs will be low, (c) management gets a chance to become acquainted with ADP without incurring any investment risks, and (d) as experience is gained, changeover to more advanced uses is readily accomplished.

Feasibility Check

The value of the system, in terms of clerical work savings, better operating information and better data for decision-making, must exceed the costs of ADP. The higher the value/cost ratio, the greater the feasibility of the system.

Estimates form the basis of feasibility evaluations. A simple point system of analysis, as in Fig. 21-3 aids in deciding whether

ADP JUSTIFICATION POINT SYSTEM

Category	Criterion		Point Value	Total Points
	Units	Amount		
Checks issued	No	____*	0.1	____
Employes	No	____	1.0	____
Active accounts	No	____	0.2	____
Invoices	No	____	0.1	____
Purchase orders	No	____	0.1	____
Items stocked	No	____	0.1	____
High-valued items	No	____	0.1	____
Strong seasonal variation?	Yes	10	1.0	____
	No	0	0	
Other special problems?	Yes	10	1.0	____
	No	0	0	
Total				

*Enter number per month

Figure 21-3. Point system for justification of feasibility study. In general, a total of 100 points will justify investigation of at least a plain batch system of automated data processing, ADP.

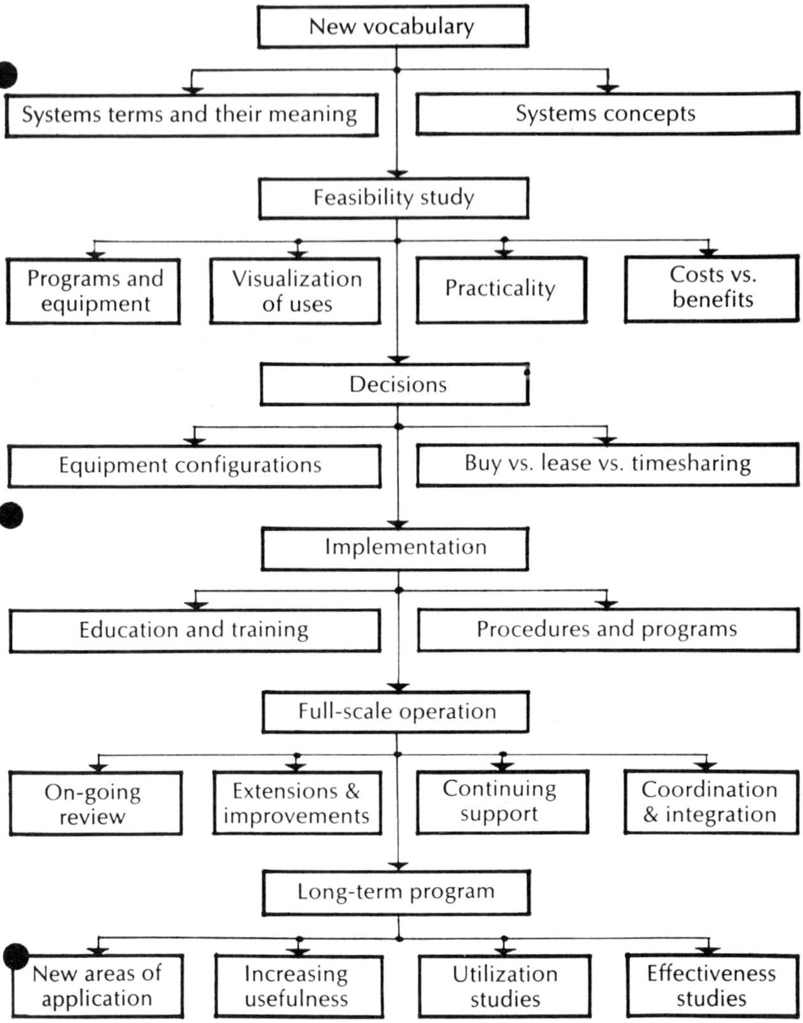

Figure 21-4. The computer information system from conception to completion.

or not a feasibility study should be undertaken. In general, if the total number of points for a particular business exceeds 100, at least batch processing of data will be justified.

The ADP justification check is, of course, just a first step, after which more detailed study is undertaken. Most feasibility analyses underestimate the values to be gained from an ADP installation. Commonly, once an information system has become operational in an organization, and executives, managers and staff people have become acquainted with its vocabulary and implications, they find new, valuable uses for improved clerical work, operating performance and short and long-range decision making. The flow chart in Fig. 21-4 illustrates the way in which the system develops.

While in many of the preceding sections a sequential reading of chapters was often not necessary, for the present section it should be emphasized that a proper appreciation of the material is possible only if each chapter is read successively, without skipping.

The major aspects of installation, operation, control, and maintenance of a management information system are presented, all with a consideration of their operations research implications, and from a viewpoint of practical applications.

CHAPTER 22

The System Feasibility Study

Introduction

A feasibility study serves to identify the principal areas in which computer information systems save costs and increase a textile company's profit opportunities. Close coordination between feasibility team members and operating management guards against impractical installations and ensures that the system chosen will deliver useful and timely information flows.

Cost, Value and Operational Feasibility

A feasibility study should always precede the installation of any sizable computer information system. Technological feasibility is a matter of expert judgment, and the status of present-day data processing equipment makes this type of feasibility widely available. This leaves three principal factors to be resolved:

1. *Cost feasibility:* Automated data processed should be less costly than manual or mechanical methods.

2. *Value feasibility:* Information generated by automated data processing should result in knowledge that permits better decisions in such areas as inventory policy, production scheduling and sales.

3. *Operational feasibility:* Company organization, human relations and political factors promote the translation of knowledge into action that improves operations, such as inventory management, production scheduling and sales. These feasibility factors are underscored in Fig. 22-1.

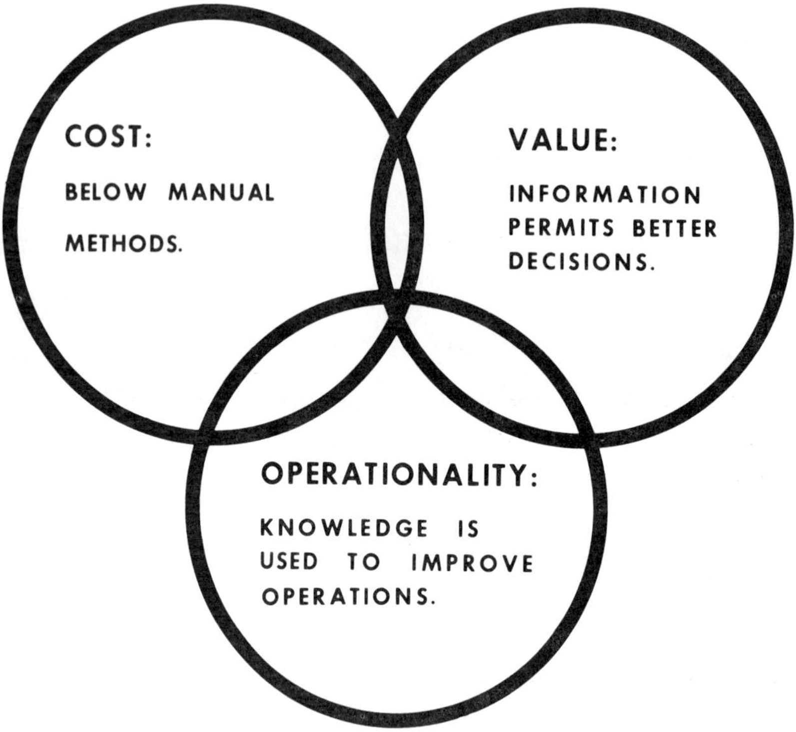

Figure 22-1. Feasibility factors. Relationship of cost, value and operationality are stressed.

Three Stages of Usage

Detailed analysis of feasibility within an organization implies an understanding of three basic stages of computer use as shown in Fig. 22-2:

1. Automation of record keeping forms the first stage of computer data processing. Efficient installations are immediately recognizable in terms of their cost feasibility. Clerical costs are reduced, speed and accuracy improved. Virtually all organizations of middle and large size rely on computers—via batch processing, time sharing or on-site operation—for their important record-keeping.

2. The second stage involves more sophisticated appli-

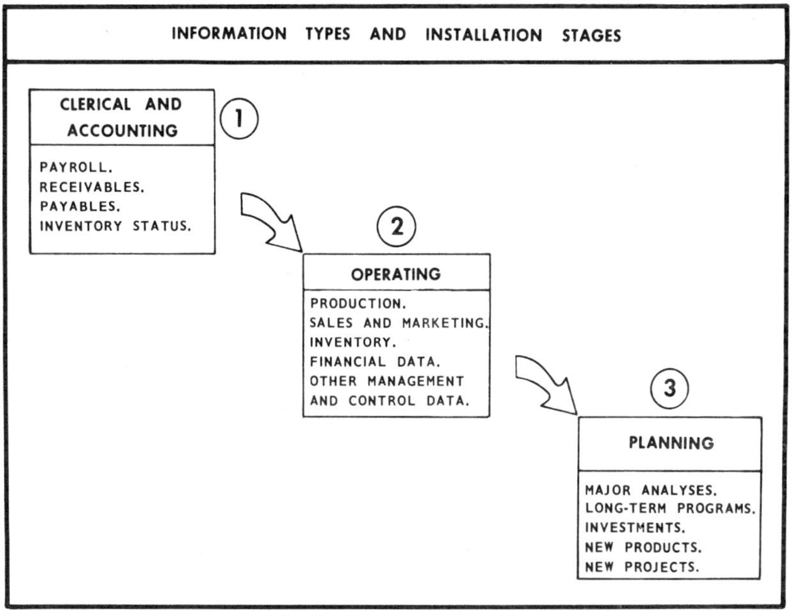

Figure 22-2. Three stages of development of Computer Information Systems.

cations: computer-aided scheduling to smooth production flow and avoid shortages or bottlenecks at various machines; or a system of short-term demand forecasting to help in the control of inventories and the avoidance of undue shortages or excessive supplies; or analysis of seasonal patterns, market and styling trends to guide promotion and sales efforts.

3. The third stage, generally reached only after a period of years, involves relatively complex analyses of information used in planning policy and strategy. Simulation of the mill's major operations, for example, checked against environmental trends and competitive developments, will yield indications of the likely effect of alternative management decisions. A product mix analysis will highlight those fabric styles that are both marketable and well-suited for balanced production. And computerized analysis may help in developing sound bid levels on a variety of contracting opportunities.

Stages and Feasibility

Cost feasibility factors are the prime consideration in the clerical and record-keeping phases of computer use. At this stage, management's main concern is to acquire appropriate hardware and install procedures that reduce expenditures. Improved speed and accuracy also are important, though usually subsidiary, considerations.

But in the second stage, value and operational feasibility become the significant considerations. We are no longer satisfied with records and reports. The information must be processed further, so as to develop results that are immediately useful in operating decisions affecting inventories, production and sales. Astute management will not confine its feasibility study to the first stage of computer application, but will look ahead to this second stage at least.

The third stage, involving complex analyses, represents a highly advanced development of the second stage. Usually, a good many years will elapse before it is reached. Since detailed long-range planning may not always be expeditious, it may well be that a good feasibility study will end somewhere in the second stage.

In general, the first stage—a good record-keeping system—can be installed within a few weeks to six months. The second stage, in which the computer is applied in the firm's operating decisions, can be accomplished within a year, but will be reaching maturity only after the second, third or fourth year. Strategic and policy planning applications—the third stage—occur after five to ten years. While great benefits may accrue to the company which energetically pursues third-stage analyses and decisions, it will need much initial investment, not only in systems study costs but also in executive time.

Feasibility Relations

Most companies will investigate computer processing feasibility for the first and second stages of application. Con-

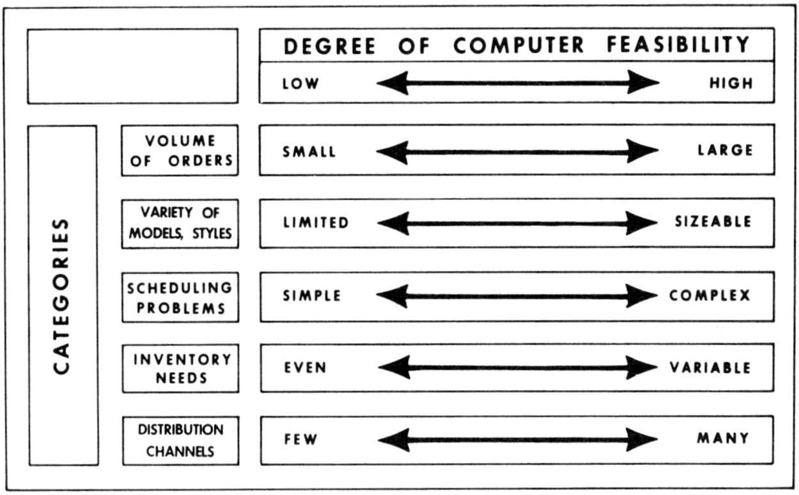

Figure 22-3. Operating conditions and feasibility. For example, highest benefits from computerization may be expected when (1) volume of orders is large, (2) variety is sizeable, (3) scheduling is complex, (4) inventory needs are variable and (5) there are many distribution channels.

sidering the types of decision problems that will be attacked, the feasibility relations of Fig. 22-3 apply.

Computerization is likely to yield the highest benefits in a company which has a large volume of orders; where there is a large variety of fabric or knit goods styles to be sold and produced; where scheduling involves many processes and materials flow arrangements; where inventory needs vary with customer demand, seasonal factors and similar considerations; and where there are numerous distribution channels. In dealing with so many alternatives and possible events, a computer, with its enormous information storage and processing capabilities, will be of considerable value in digesting all the inputs and producing results that aid decision-making.

Conversely, where orders are few, models and styles limited, production lines straightforward, inventory needs well-defined and distribution channels fixed, there is less need for computer information processing, less need for intense analysis of multiple interacting factors.

Steps in a Feasibility Study

Let us say that a textile company has set up a feasibility study team consisting of two systems analysts, two representatives from production, two from sales and one from accounting. A possible program is outlined in Fig. 22-4. Let us look at the rationale.

1. The team task is defined, which calls for determining (a) the extent of operating management's requirements of information for decision making, planning and control; (b) economical and practical means of meeting the requirements; (c) ways to coordinate data processing with supplier and customer systems; (d) suitable hardware configurations and software packages; (e) organizational structures promoting effective use of data processing results, and (f) installation priorities and time schedules.

2. The team acquaints operating management with the potentiality of computer information systems. As part of this program, computer simulation games may be played, in which each group of three or four participating executives forms a "company," which then "competes" against the others. Production, marketing and financial decisions must be made in rapid order. Within a single day, some 20 decision cycles will have been completed, and the most effective group identified. All participants learn to appreciate the speed of the computer in processing the decision inputs and providing outputs in terms of inventories, costs and profits, as well as comparative performance charts. Usually, these simulation games make it easier for the study team to elicit from operating management the essential information on its decision-data needs. The operational effectiveness of whatever system may be adopted will depend on its ability to provide essential, useful and timely information flows.

3. Some quick projections of anticipated costs and benefits are needed. Only if calculations indicate an excess of gains over costs can computerized information processing be justified.

4. Attention is turned to priorities: which projects are most urgently required in the light of data processing needs?

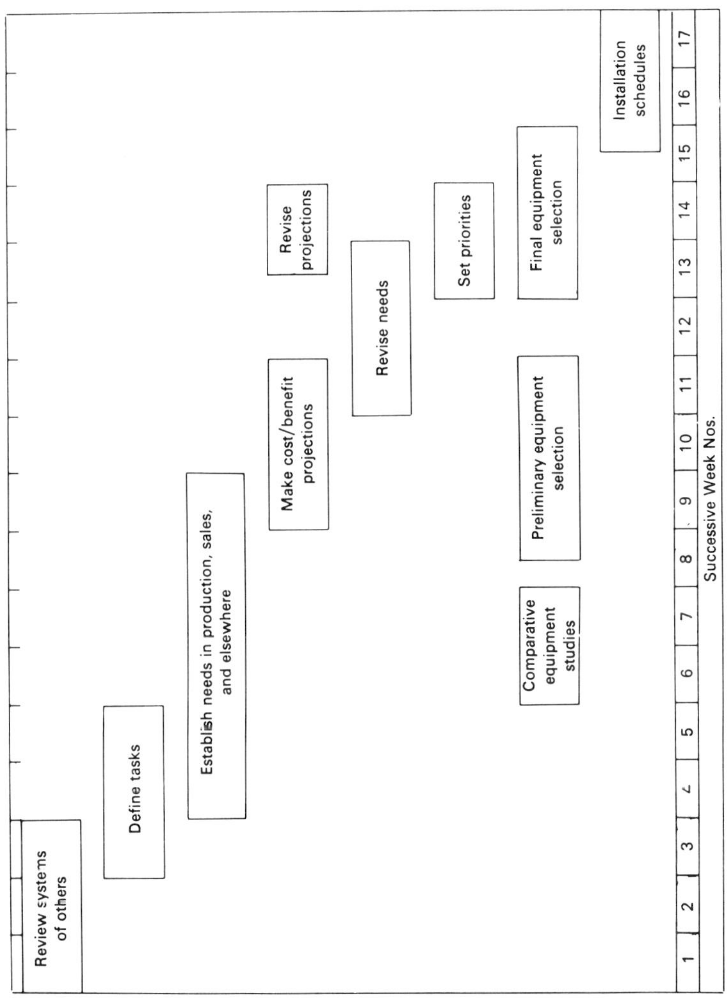

Figure 22-4. Time schedule for making feasibility study.

These priorities in turn become the basis for installation time schedules. The more advanced applications will necessarily be scheduled for installation sometime after the basic record-keeping systems have become operational. The data bases developed for the earlier, simpler applications become the building blocks for more sophisticated programs.

Fortunately, the feasibility team does not have to start entirely from scratch. Information is available about the problems, procedures and achievements of feasibility groups from other companies in the textile and related industries. Some of this material is available from published case histories. In other instances, personal visits may have to be made to companies experienced in various areas of programmings of MIS applications.

CHAPTER 23

The System Cost-Benefit Evaluation

Introduction

Cost-benefit analyses relate expenditures to expected gains from the installation and operation of automated data processing (ADP) systems. Such analyses are needed to justify new investment or the enlargement of existing facilities and programs.

Cost-benefit data provide a goal for attainment: Costs must be kept within reasonable limits, and the information system should be developed and maintained in a way that ensures a flow of information of adequate value in management for decision and control activities.

General

Whatever the type of automated information system, its costs require justification in terms of the benefits attained. Cost-benefit evaluations become mandatory in these types of situations:

1. During the design of a new system, as part of the feasibility study, we need projections of future expenditures and monetary as well as intangible benefits.
2. When revamping or enlarging an existing installation, we must ascertain whether the information improvement is great enough to warrant the costs involved.
3. At the occasion of periodic reviews by mill management of the efficiency of computer utilization, ratios

of costs to benefits are major yardsticks of accomplishment and performance.

Cost-benefit evaluation achieves more than the justification of expenditures: It also illuminates specific areas where operating results and accomplishments are inadequate. Special management attention and corrective efforts can then be concentrated in these areas.

Types of Benefits

Benefits bestowed by the information system are:

1. Savings in accounting, clerical and record-keeping costs, as a result of automated, systematized operations. . . . . .
2. Increased profits, resulting from better decisions, aided by timely, relevant and well-digested information for production, marketing and financial management.
3. Enhanced strategic planning and policy development, based on mathematically founded analyses of relevant data.

The two principal types of computer system benefits, systems operation savings and information-generated gains, are portrayed in Fig. 23-1. Reduced costs of clerical, accounting and record-keeping functions are readily expressed in monetary units. Information gains are also convertible to money terms. For example, improved knowledge on anticipated market demand by fabric styles may permit a company to gear its inventory build-up more closely to customer needs. As a result, the amount of "distressed" merchandise left over at the end of the season is minimized. The amount of money saved thereby is an information-generated benefit. As a further example, computer-aided mill processing scheduling may achieve smoother production flow with fewer waiting lines and idle machines. The consequent labor and machine-time savings, as well as the production improvements, represent a further instance of information-generated benefits.

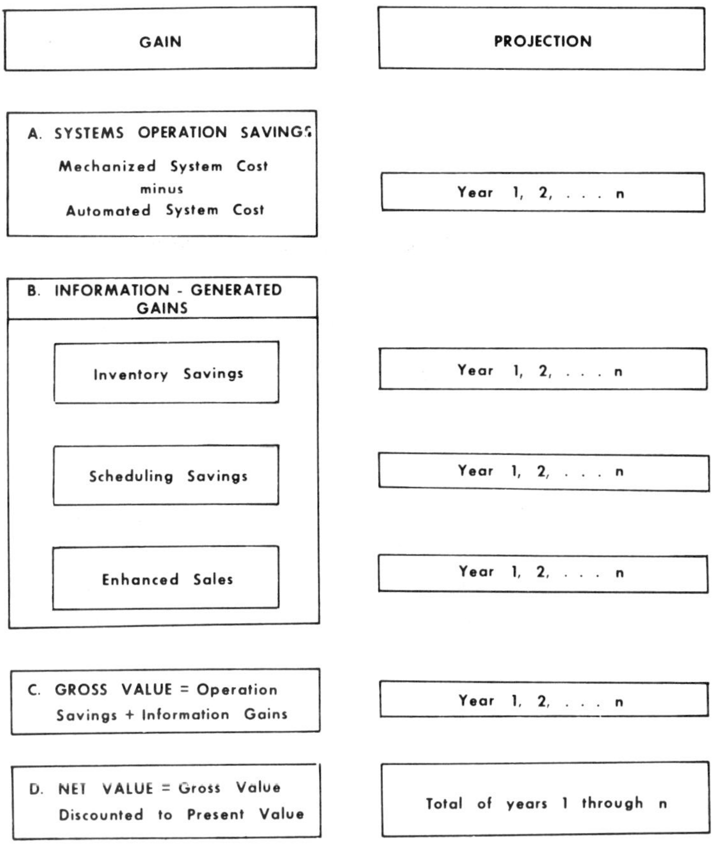

Figure 23-1. Integrated data processing benefits.

Benefit Projections

A new installation or revamping of an existing system requires considerable initial expenditures. It would be erroneous, therefore, simply to project expenses and gains over a number of years and come up with the equation:

$$\text{Gross Value of System} = \text{Systems Operation Savings} + \text{Information-Generated Savings}$$

all converted to monetary units.

We say that this approach is deficient or incorrect, since it does not give consideration to the need to discount future earnings to present value. By "discounting" we refer to a process of allowing for the cost of borrowing money or, alternatively, for the interest that we might have received on our own cash, had we lent it out. The prevailing interest rate may represent a useful indicator of the cost of money.

Benefits from operating savings and information-generated gains are thus best expressed as:

Net Value = Gross Value Discounted to Present Value

with "Gross Value" determined as shown in the preceding equation.

The accuracy of the net value figure will depend on the quality of the forecast. Discounted present values are readily obtained from tables, using the interest rate figure chosen by management as most applicable. It should be understood that anticipated gains from systems installations or expansions are fully realized only in those instances where senior and operating management remain actively involved in the ADP program.

Illustrative Example

Cost-benefit projects for a computer information system (CIS) appear in Table 23-1, with the following major features:

1. The difference between mechanized and automated system costs represents the savings from automation. The investment of $56,000 is carried as a "loss."
2. Benefits generated by better information are expressed in terms of improved inventories, production schedules and customer relations.
3. Projections are made for four years. A range from four to six years is customary. It is difficult to make projections too far in the future.
4. In converting gross savings to net benefits, management chose a relatively low interest rate factor of 6 percent.

TABLE 23-1
Cost-Benefit Projection for Computer Information System
(All Entries are in Thousands of Dollars)

Category	Invest-ment	Year 1	2	3	4	Total
Mechanized System Costs						
Staffing		18	20	22	25	
Equipment		2	3	4	5	
Supplies		1	2	3	4	
a. Total		21	25	29	34	
Automated System Costs						
Staffing	50	6	7	8	9	
Equipment	15	10	11	12	13	
Supplies	1	1	2	3	4	
b. Total	56	17	20	23	26	
Automated Systems Benefits						
c. Systems cost savings (a-b)	(–56)	4	5	6	8	
d. Inventory (lower stocks through better coordination of sales with production schedules and inventory)		3	6	10	11	
e. Scheduling benefits (Smoother work flow, less idle time, fewer parts shortages, minimized waiting)		2	3	5	5	
f. Customer relations (more reliable delivery, better customer information service, increased goodwill)		1	2	4	4	
g. Gross savings (c+d+e+f)	(–56)	10	16	25	28	
h. Present-value factor at 6 per cent interest rate		.94	.89	.84	.79	
i. Net savings benefit (hxg)	(–56)	9	15	21	22	10

Note: Present value factor is computed from $1/(1 + i)^n$, where i is the interest rate (6 percent in our example) and n is the number of years. Thus, in year 2, factor is $1/(1 + 0.06)^2 = .89$ (rounded).

Rates from 8 to 15 percent are more usual. Although we have shown the formula for present value factor, in practice one looks this up in tables.

The cost-benefit analysis projects a present value of $10,000 against an investment of $56,000. Ordinarily, one might not view such an investment as attractive. The system is expected, however, to operate well beyond the first four years analyzed, and further benefits will accrue. Moreover, as people work with a computerized information system, they inevitably discover and develop new, expanded applications of value to decision making.

The projections assume continuing active involvement on the part of management. As in all ventures, the degree of ultimate success is not guaranteed. It is governed by the effort expended.

TABLE 23-2
Illustrative Annual Costs of Medium-Sized Systems Operation
(Time Sharing Installation)

Account Category	Costs, $1,000 Item	Costs, $1,000 Sub-Total	Per Cent of Grand Total
Salaries and wages	130		65.0
Employee benefits	15		7.4
Training	10		5.0
General allocations	5		2.5
Total personnel expenses		160	80.0
Space rental	10		5.0
Equipment rental	20		10.0
Depreciation	2		1.0
Taxes	2		1.0
Total occupancy and use expenses		34	17.0
Mailing, stationary, forms	2		1.0
Telephone, telegraph	1		0.5
Travel	1		0.5
Insurance	1		0.5
Miscellaneous	1		0.5
Total other direct expenses		6	3.0
Grand Total		200	100.0

Note: This medium-sized application pertains to a relatively mature system for a firm with annual sales of $20 million. All entries modified, simplified and rounded for illustrative purposes.

Cost Relations

We have illustrated an example in which equipment costs (time-sharing rentals and ownership of some peripheral equipment) exceed salaries. Such a relationship is not always the case. Much depends on the type of equipment utilized, the proportion of ownership versus rental, and the kinds of programs included in the system. Table 23-2 brings a further case history, in which salaries and wages exceed the rental for space and equipment and depreciation on owned facilities.

CHAPTER 24

The Use of Models

Introduction

When management programs and task situations involve a great number of variables and interlocking elements, analytical models help in assessing the major effects of the predominant factors at work. Values and relationships are laid down definitively in the form of mathematical equations, that mirror the real-world relationships.

After management is satisfied with the evaluations so obtained, has reviewed the various alternative actions and anticipated outcomes, and has weighed both short-term and long-term considerations, an appropriate course of action is chosen.

While computer processing is essential in the evaluation of various alternative courses of action, the human task of building relevant models of the problem situation must precede the numerical analyses. This process, when followed in a step-by-step manner, using cooperative endeavors of manager and analyst, will be straightforward. The actual task, as will be seen, is not a difficult one.

General

A major aspect of decision processes in the textile industry, from raw materials purchases through manufacturing, marketing, distribution and sales, is the evaluation of alternative courses of action, within the confines of various limiting factors, in the search for optimal courses of action. Models help in this decision making. They apply to any complex system and in essence represent a small-scale replica of real-world relations pertaining to a decision problem. Usually, a model is comprised of a set of key variables and a system of equations defining linkages and restrictive factors. Inevitably, some degree of approximation and simplification is required when seeking to mirror, in terms of mathematical formulations, the immensely complex and intricately ramified aspects of an actual problem situation.

OFF THE SHELF MODELS – EXAMPLE OF APPLICATION in MARKETING

1. State the problem.
   We wish to determine profit contribution for a marketing project.

2. Find a source.
   An accounting text will serve.

3. Define constants and variables.
   There are no constants. Variables, in dollars per time period, are profit contribution C, sales S, production cost P, distribution cost D and marketing cost M.

4. Write the equations of relationship.
   C = S – (P + D + M)

5. Consider refinements.
   The model implies linear relations. For the range of values of interest it will not be necessary to do curvilinear analyses.

6. Validate the model.
   The model is logical and well established. We will merely check for reasonable answers. One check shows:
   C = 50 –(20 + 15 + 10)
   = in $1,000 or $5,000
   This looks reasonable.

7. Confirm the model.
   The model does provide the needed profit contribution C for various values of S, P. D. and M.

Figure 24-1. Marketing model. Off-the-shelf development.

Off-the-Shelf Models

Modern mill management has many uses for models that already exist and can thus be taken "off the shelf." An organization's files, procedure manuals, computer software, reference texts, and formulas appearing in handbooks or current periodical literature constitute some of the numerous sources of off-the-shelf materials. It would be pointless to re-develop existing models.

Steps in the application of existing models are: (1) state the problem; (2) find a source or sources; (3) define constants and variables; (4) write equations of relationship; (5) consider refinements; (6) validate the model; and (7) confirm the model. We illustrate these concepts with a marketing example in Fig. 24-1.

Scientific aspects of model building are represented by the use of formulas and relationships which have become a part of general knowledge, generated through numerous studies, experiments, and rigorous logic. But there are many aspects of art in the use of this material; observing a current problem, recognizing a properly applicable model, and using judgment in deciding on the degree of model refinement required.

New Models

New models are developed when management and staff recognize the need for an analytical approach in dealing with a major project, program, task, or problem situation. Close cooperation among management, operating, and mathematical-analytical personnel is prerequisite to effective model development. Moreover, a particular model is rarely fully "final." After implementing the model, modifications, adaptations and revisions will be required to conform to new economic, political, marketing or other trends. In general, insights gained from model building experience should help in the evolving of analytical structures of high validity and usefulness.

Development Sequence

Specifically, development, use, and revision of a model calls for a series of distinct steps, as illustrated in Fig. 24-2. These are elaborated on below.

1. *Identification of Problem.* Operating and staff people concerned recognize a real-world problem. Significant variables, relationships, and interlocks are identified.

2. *Review of Alternatives.* Although a number of alternative problem solutions become apparent, there should nevertheless be a search for all feasible approaches. Inherent in this process is a review of the alternatives in light of the restrictions, limiting factors, and drawbacks that apply to each.

3. *Compilation of Data.* Information on volumes, costs, prices, expected gains and losses, and the like, as well as the basic relationships among these factors, constitute the materials from which analytical formulations and solutions are derived. Much of what is needed will be found in data banks or various other files. Where gaps exist, additional information must be secured. Estimates, based on brief sampling surveys, will often turn out to be acceptable substitutes for costly full-scale studies.

4. *Building the Model.* Practical operating managers and staff people work with mathematically and computer-oriented analysts to review the important factors and relationships of a problem and express them in the form of a comprehensive model. The model thus is a marriage of practical operating and sophisticated mathematical expertise.

5. *Formulating the Solution.* The mathematical-computer analyst searches for a means to solve the equations relevant. He may develop maximizing or minimizing functions for the factors of interest; he may find a set of optimal decision points or ranges; or he may define a systems equilibrium. These activities depend on the nature of the problem involved.

6. *Computer Analysis.* The solution is usually obtained from a large number of calculations, thus requiring computer programming and processing.

7. *Solution Outputs.* The final computer outputs represent the problem solutions. It is necessary to validate whether

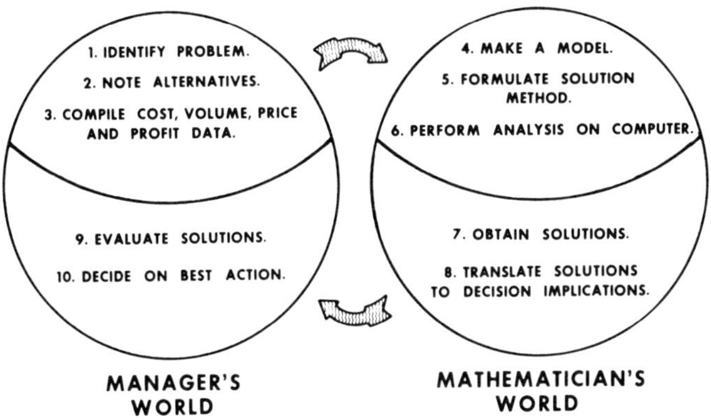

Figure 24-2. Interaction of manager and mathematical analyst in model building process.

the solutions correspond to the original intent. Do they address themselves to the problem questions posed? Here, as well as in all prior stages, a return to earlier steps is required whenever it is discovered that major objectives are not satisfied.

8. *Translation of Solutions.* Mathematical solutions may be difficult to interpret directly. They must thus be transformed into their practical meanings and implications. This interpretation process includes condensing the findings into useable form, with a clear overview of the principal results. For example, several major alternative actions may have been found profitable. For each the expected gains, costs, and risks should be shown clearly.

9. *Evaluation of Solutions.* Management now reviews the solutions in terms of their short, intermediate and long range effects. Checks are made for consistency of consequences with major policies. Involvement of major factors of marketing potentials, production facilities, and likely competitive departments, as well as general economic and other trends, are given careful consideration. Should it be found that important categories have been underestimated, revisions of original assumptions and renewed analyses will be needed.

10. *Decision.* Whatever course of action management chooses will represent the best decision in the light of available, analyzed information. Each action, in turn, brings new problems and new opportunities in the area of operations, marketing and service. Eventually, revised formulations, new models, and new analyses will be needed to deal effectively with these and other developments.

Model building is thus a never-ending program, particularly since improvements in one area of operations soon lead to opportunities for similar benefits elsewhere, thus enhancing the overall performance of the organization.

Mathematics of the Model

Thus far we have discussed model development within the overall context of organizational requirements and interaction among the persons responsible. There are also recommendations for structuring the model in a mathematical context,

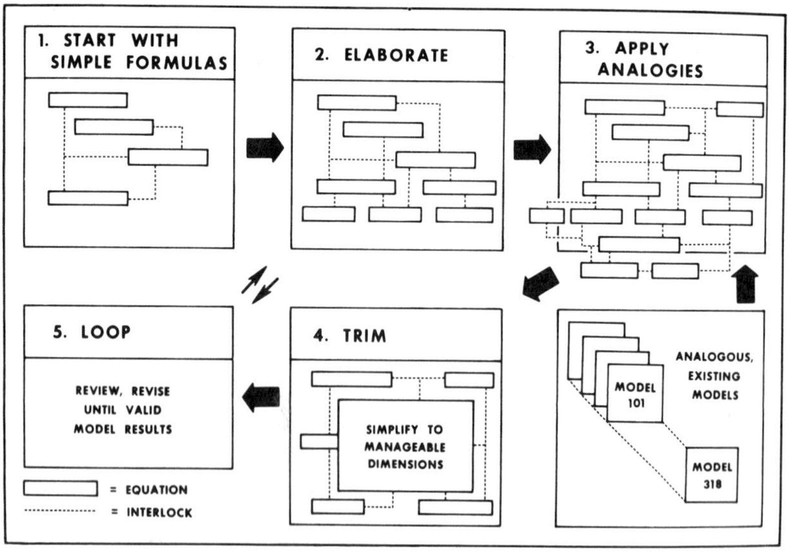

Figure 24-3. Mathematical steps in model formulation.

relating to the variables, equations, and interconnections that must be expressed in useful form. It involves a search-exploration as laid out in Fig. 24-3.

We begin with known relationships and then elaborate by reference to further variables, equations, and interconnections. A viable model begins to emerge. Usually, however, the result will be incomplete. Reference to similar, parallel or analogous

MODEL SIMPLIFICATION GUIDELINES

* Constants for variables.
  Narrow-range variables may be replaceable by constants without harming validity.

* Eliminate variables.
  Omit variables which, by means of trial runs or other evaluations, are shown to be insignificant.

* Minimize on equations.
  There will be many equations of relationship. Some may be of little over-all consequence. Restrictive or limiting equations often fall in this category. They may be dispensable.

* Use main alternatives.
  Not all alternative actions can be considered. Exceptional or remote items should be treated outside the general model.

* Linear approximations.
  Curvilinear relationships are hard to handle. When curves are not highly marked they may be approximated by a straight line. At other times, two or three linear segments can approximate a curve.

* Strengthen main structure.
  At all times, concentrate on the main structure. It is not feasible to consider all detail. Omit items of minor impact.

Caution: Apply simplifications with care, judgment and circumspection.
We need a model of manageable, tractable and comprehensible dimensions.
In case of doubt use trial evaluations, to assure that significant variables are not ignored.

Figure 24-4. Trimming the model to manageable size.

structures from existing models will give clues to refinements. During this process, the model may become too sizeable, and may have to be trimmed to less ambitious dimensions from both a data-processing and managerial comprehension viewpoint. Fig. 24-4 provides simplification guides.

Feedback loops, which serve to constantly review the progress of work, check for omissions, and consider alternative variables, equations, and formulations, are integral to the revising and refining process.

Model-Building Skills

Rarely is a model entirely new. Rather, it will be constructed from older ones by using creative combinations and extensions of concepts and relationships. Modeling skills are gained gradually. Significant elements in this process are:

- Experience with the application of off-the-shelf models, from which are derived the heuristic abilities to adapt existing formulations to specific situations.
- Cooperative spirit among managerial and analytical personnel, aimed at isolating variables and serving to encourage the exploration of new approaches.
- Confidence developed as success with model usage is experienced.

Time and experience thus sharpen an organization's ability to develop and utilize models effectively.

Illustrative Case

We will illustrate the process of model building with the aid of a marketing task. This starts with the basic formulation given previously but introduces further information on past experience (see 24-5), with the objective of making forecasts for one or more future years.

For each year, the profit contribution of each product is shown by various market areas in the form of a matrix. It is thus possible to make backward comparisons for each of the

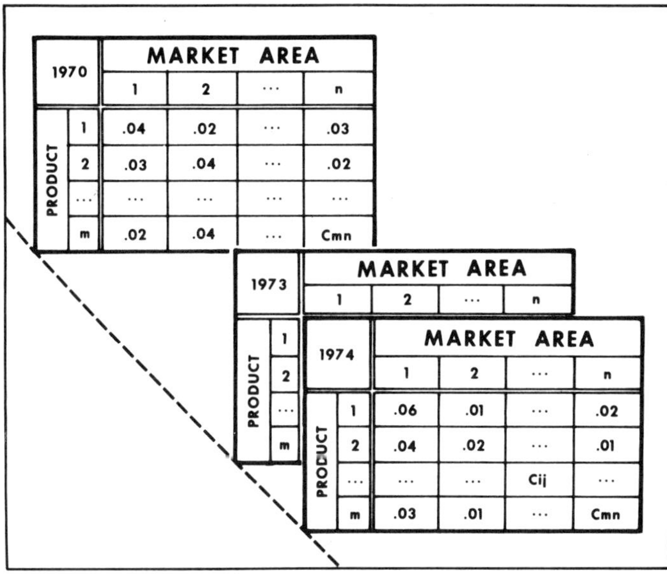

Figure 24-5. Historical array of product and marketing area matrices. Profit contributions C_{ij} are shown for m products in n markets. C_{ij} is the contribution of product i in market j.

cells of the matrix presentations. Each matrix also is an extension of our earlier off-the-shelf model:

$$C = X - (P + M + D)$$

where symbols have been explained previously. Further, if we express the costs, P, M and D as a decimal fraction of sales revenue S, we obtain for each profit contribution entry C_{ij} the result

$$C_{ij} = 1.0 - (P_{ij} + M_{ij} + D_{ij})$$

and a new set of matrix entries.

Decision Problem

Prime interest centers on the decision problem for the next marketing period. For this purpose, past matrix values for several years constitute a background of experience and trends.

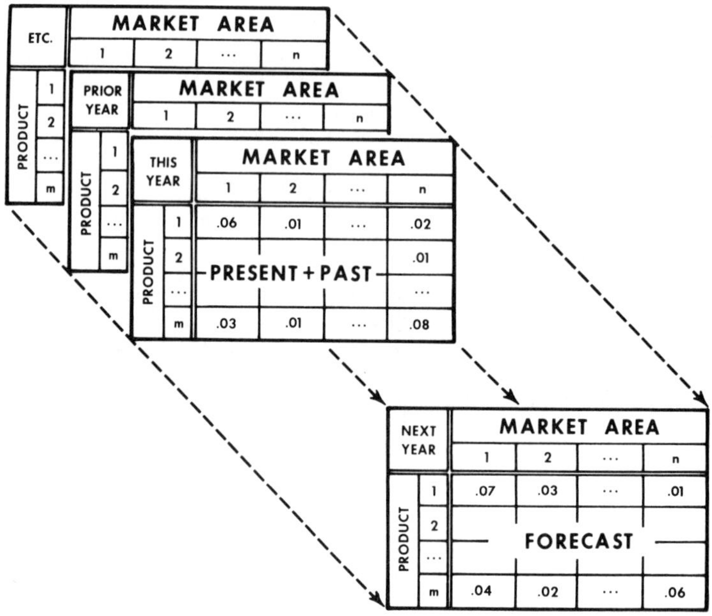

Figure 24-6. From historical array, forecasts are obtained by projection and related techniques.

Least-squares trend lines may be developed and extrapolated, as indicated schematically in Fig. 24-6. Anticipated contribution, C' is now found for each cell in terms of

$$C'_{ij} = 1.0 - (P'_{ij} + M'_{ij} + D'_{ij}).$$

Initial decisions are now feasible. We can, for example, rank the C'_{ij} values in descending order, reflecting relative short-term profitabilities. By reviewing these and other forecasts, considering requirements for sound overall marketing policy, and allowing for production factors, management can allocate appropriate manpower and funds for selective product promotions. Additional models from mathematical programming, distribution logistics and other analysis methods may now be applied for a further quantitative review of key variables in management's decision making. The ultimate outcome will appear in a decision matrix for the next year, following a sequence of procedures as given in Fig. 24-7.

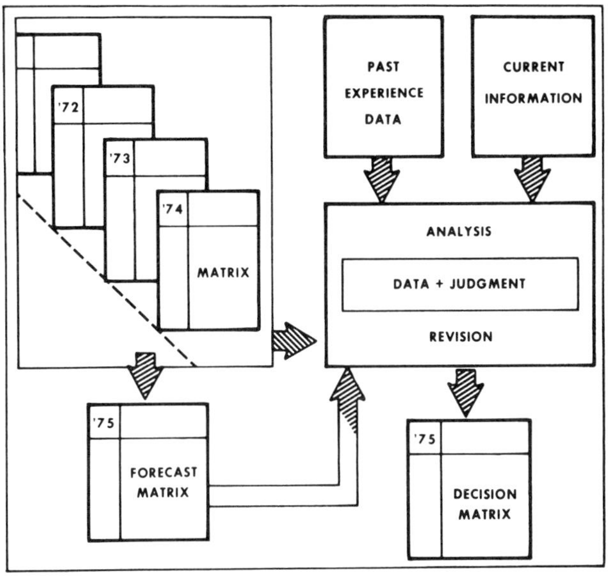

Figure 24-7. Decision matrix sequence. From past experience to forecast, to management evaluation and ultimate decision matrix. The decision matrix specifies the plans and schedules for the next period.

Refinements

The anticipated profit contribution and cost ratios in our example were generated by extrapolation from past data. Uncertainty attaches to such values, and a more sophisticated Model takes account of this by:

- Providing for each product and market area a range of likely sales volumes and profit contributions, together with relevant probability values.
- Showing indexes that incorporate a composite of management judgment regarding the probable occurrence of exogenous events, such as competitors' counteractions,

FEATURES GOVERNING THE QUALITY OF MATHEMATICAL MODELS

Clarity

Variables and equations are defined and explained in a clear-cut manner. Management and staff understand and can properly interpret the major outputs.

Completeness

All major relationships have been included.

Practicality

There are no major difficulties in data inputs. Outputs of results have a direct bearing on the real-world problems and relationships involved.

Testability

Provisions have been made for quick tests of validity of the model and its outputs. Where no formal tests are possible, the data should be in such a form that intuitive understanding and confirmation by management and staff is possible.

Value

Application of the model should yield data, recommendations and conclusions that are of immediate usefulness in managerial decision processes. The more useful results, the greater will be the value of the model.

Fertitility

The model should facilitate eventual further developments of additional models dealing with problems in related areas of management concern.

Flexibility

In a world of accelerated economic, political, technical and marketing changes, the model should be easily adaptable. Modifications based on new insights or new outside developments will be needed frequently, and it should be clear that the model must be so designed as to accommodate these revisions readily.

Robustness

Design of the model must be with a view to stability. The model should not be unduly sensitive to small changes in the assumptions that characterize it.

Versatility

Many decision problems may have to be answered by one major model. It is thus important to build versatility, in terms of a variety of decision-serving information outputs, into the formulations.

Figure 24-8. Criteria for evaluation of model quality.

effects of likely government policies, and foreign imports.

The model becomes more complex, but also more specifically tailored to most of management's planning requirements. Model building represents a growth process in which value and usefulness are enhanced gradually. At each stage, quality of the model is related to the features given in Fig. 24-8.

Summary

Practical, useful planning models are a function of the skill and ingenuity applied during their development. A good deal of creative art, within a framework of systematic, scientific procedure, goes into the building of new marketing models. In practice a mill organization may do well to gain experience with off-the-shelf models and to proceed to specific, tailor-made projects gradually as experience and skills grow. Guidelines for overall programs and specific procedures in model building are provided and illustrated with examples.

CHAPTER 25

Systems Simulation

Introduction

Time was when a logical, sound-appearing management policy could be put in effect with little advance testing. "If it does not work out, we can always make a change," was the attitude. This approach is no longer feasible. Competition, technological acceleration and time constraints make the "right decision the first time" the important goal that must be followed.

Computer simulation is a technique which permits management to try out dozens of options "on paper," reviewing the outcomes in terms of computer outputs, and selecting that set of decisions, plans or policies which seems best suited to achieve desired goals. While simulation does not guarantee success, it nevertheless is a means of attaining greater insights and perspectives than unaided judgment.

We may view computer simulation as a "time compression" technique, in which the effect of numerous variables is tested quickly, thus providing an "advance look" at likely outcomes from various decision alternatives. It is assumed, of course, that the decision problem itself can be and has been expressed in the form of a model (as discussed in the prior chapter). While models can often be solved by purely mathematical analysis, large-scale or complex problems are more readily handled by simulation techniques.

Procedures with Investment Example

In general, simulation represents a form of imitation of real-life relationships. Computer simulation is applicable to practically any process, such as production, marketing and financial decision problems. In particular, simulation involves a series of analytical steps, as described below:

1. A problem is noted. For example, management may have to decide on the size of a new finishing plant from a viewpoint of long-term profitability.

2. Using available data, extrapolations and estimates, relevant relationships are postulated, such as for price and sales volume, size of investment needed to support various sales volumes, and expected degree of market penetration at various price levels over a period of years.

3. A Model is prepared in equation form, showing how all of the individual relations work together to yield total cost, total revenue and profit.

4. A first simulation trial is run for one pricing policy and sales volume schedule, one plant size, one degree of market penetration, and one level each for everyone of all the further factors considered.

5. Simulation trials are repeated for all important combinations of factors and levels. Several hundred runs will usually be required.

6. Outcomes, such as costs, revenues and profits, are observed. Faced with a problem of plant size investment, management will probably be most interested in long-term profitability.

7. The decisions (such as plant investment and pricing policy) that are likely to yield highest gain are identified. Long-term profitability would be the criterion in our example.

Once a system or model of the equations of relationship has been set up and the programming has been completed, the actual computer run of hundreds of trials is accomplished in a day or less. The result is an advance look at the likely long-range effects of various decision alternatives. Guided by these perspectives and insights, management will be able to choose the most logical course of action. Obviously, not just computer runs, but also qualitative and judgmental factors that cannot be put in quantitative form, will play a role in the decision process.

A schematic of the simulation process is illustrated in Fig. 25-1. Obviously, the plant size investment decision is of a long-term nature, while other decisions, such as pricing policies, are more flexible.

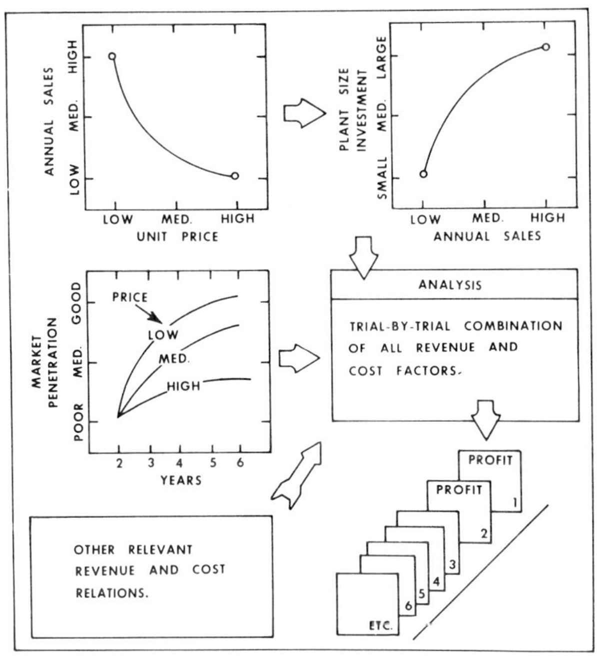

Figure 25-1. Simulation of investment and pricing problem.

Further Applications, Quality/Cost Example[1]

Management policy and strategy are not the only instances in which simulation is of value. Practically all levels of operation of an organization lend themselves to some aspects of useful simulation studies. The consequences of smaller-scale operating

[1] Adapted from Lester, Ronald H., Norbert L. Enrick, and Harry G. Mottley, Jr., *Quality Control for Profit* (New York: Industrial Press, 1977).

decisions are less crucial and far-reaching than major policy determinations, such as on plant size investment. Nevertheless, one should always look for the best information on which to base actions.

A pivotal consideration in many routine, small-scale simulations involves cost/quality factors. For example, a mill had traced unduly high weaving costs to an excessive flow of defective yarn. These problems were not discovered until final inspection, when the fabric had to be rejected. It was thought that a good amount of in-process inspection during yarn production would minimize these problems, since bad work would be discovered before it could follow through the processing stream into weaving.

Time, labor and space considerations made it impossible to perform 100 percent inspection throughout all processing stages. Sampling was thus needed. For example, a lot consisting of 200 cans of silver, would be subjected to sampling tests. If the sample found the lot of acceptable quality, it was passed to the next process. Otherwise, indications of defective product would cause rejection to screen out unacceptable parts.

Simulation was required to balance two types of cost: (1) the cost of inspection and testing labor, which varies with the amount of sampling done on each lot; and (2) the cost of rejected or second-quality final product. It is obvious that increased sampling reduces the flow of bad yarn, thus reducing later troubles, but the price paid is increasing inspection and testing labor.

The simulation model in Fig. 25-2 begins with a so-called "cumulative distribution curve," showing the quality performance of the production process. Further detail appears in Table 25-1. The sampling plans represent increasing amounts of sampling inspection and testing. The larger the sample size, the less likely it is that a lot of unacceptably high proportion of defectives will slip through. Based on mathematical probability considerations, we can make accurate predictions of the likely proportion of defectives that will slip through to the next processing step.

That processing stage will have any one of a number of

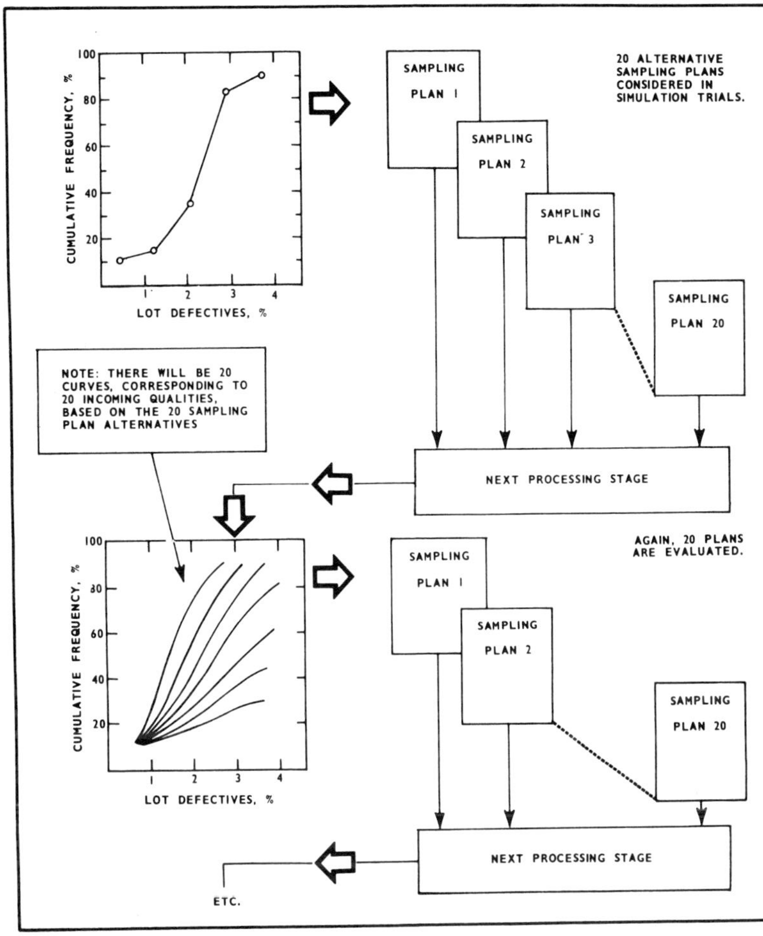

Figure 25-2. Simulation model for quality-cost problem.

cumulative frequencies, depending on (1) the defectives inherited from the prior process, which depends in part on the amount of sampling inspection done (to find and remove defectives prior to passing on) and (2) the quality of this new processing stage, which is likely to at least produce some defectives of its own.

We continue our analysis from one processing stage to the next, each time assuming a range of 20 sampling plans. Track is kept of (1) defects flow under these plans and (2) inspection and testing costs and costs of defective quality passing through processing to final weaving. For the third process, for example, the 20 sampling plans would result in the cost matrix in Fig. 25-3. Each cost C_{ij} depends on the amount of sampling after the first processing stage (i) and the amount of sampling after the second processing stage (j), plus a fixed allowance for the normal defectives experience in the third process. After sampling, of course, some substandard quality product will have been removed from the third process, a factor which will have to be considered for the fourth process matrix of costs.

Ultimately, by comparing total sampling inspection costs against the cost of defective product in advanced processing stages, a decision can be made to assure that quality (proportion of defectives) that will minimize total costs (inspection cost plus losses from rejections, repairs and second quality goods).

TABLE 25-1
QUALITY PERFORMANCE OF PRODUCTION PROCESS
Frequency Distribution of Defectives

(1)	(2)	(3)	(4)
Lot Quality, Percent Defective		Frequency of Occurrence, Percent	Cumulative Frequency, Percent
From	To		
0	1.0	5	5
1.1	2.0	25	30
2.1	3.0	65	90
3.1	4.0	10	100

Notes:
Columns (1) to (3) are derived from past processing performance. Column (4) is an upward accumulation of column (3). Thus, the first entry of 5, added to the second line of 25 gives the new value 30 percent. The latter, added to the third line entry of 65 yields 90 percent.

COST MATRIX					
SECOND PROCESS SAMPLING PLAN	FIRST SAMPLING PLAN				
	NO. 1	NO. 2	NO. 3	...	NO. 20
NO. 1	COST, $C_{1,1}$	$C_{1,2}$			
NO. 2	$C_{2,1}$	$C_{2,1}$			
NO. 3			$C_{3,1}$		
...				$C_{i,i}$	
NO. 20					$C_{20,20}$

Figure 25-3. Cost matrix showing combined effects of two sets of sampling plans, for first and second production process.

Randomness Simulator

Our calculations were based on averages. In practice, of course, there are fluctuations. In our example, lot quality and sampling effectiveness will have an average plus variation around the average. The laws of chance, in terms of "randomness' govern particular, individual outcomes, even though in the long run an average holds.

We may readily simulate variations, such as the frequency distribution of Table 25-1, in terms of a random numbers simulator, as in Fig. 25-4. The segments of the wheel conform to the proportions of the frequency distribution pattern and the respective midpoints of each group of defects. For example, lot quality corresponding to 0 — 0.99 percent defectives, with a midpoint of 0.5, represents five percent of the total of 100 percent of possible occurrences (or frequences). Midpoint 2.5, as a further example, should contain 65 percent of the total. Let us now say that the pointer on the wheel is spun around

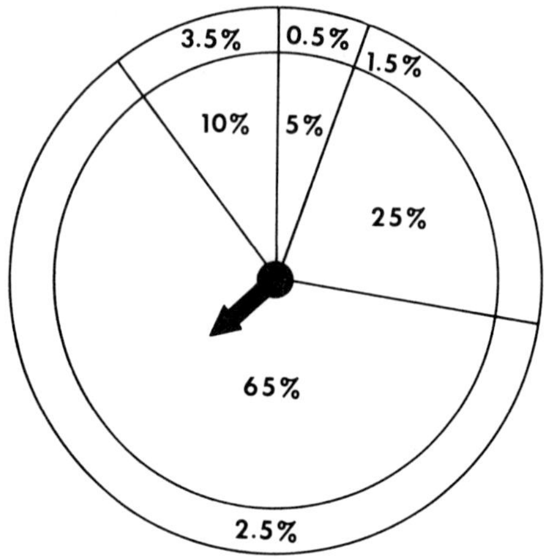

Figure 25-4. Generation of a ramdom process by means of simulation.

at random, the particular lot quality that will occur, as shown on the rim of the wheel, cannot be predicted. We can, however, expect that in the long run the value 2.5 percent defectives will come up 65 percent of the time, while for the value 0.5 percent the frequency will be 5 percent.

In practice, a roulette wheel is unsatisfactory. Any lack of precision in dimensions, bearings, and the like, would produce bias. We prefer a random numbers table. In the present instance, we would attach the following lot qualities to successive random numbers, obtained in sequence from the table:

Number Obtained from Random Numbers Table	Lot Quality, Percent Defective, Assigned
000 to 049	0.5
050 to 299	1.5
300 to 899	2.5
900 to 999	3.0

The range of numbers corresponds to the frequencies of occurrence of 5, 30, 90, and 100 percent cumulatively, given in the last column of our tabulation of lot quality. For each group, the quality midpoints represent average lot quality to be expected.

Assume now that the first random number found is 673. From the range of values just tabulated, it is apparent that we assume the lot contains 2.5 percent defectives. Let us now say our next random number comes up at, say, 041. Lot quality is now assigned the value 0.5 percent defectives.

It is a relatively simple matter to place random numbers into the storage system of a computer, or to assign a routine whereby the computer determines a random number by its own methods. Such generation of random numbers, moreover, is an essential part of practically all larger-scale simulations. Only in this manner can we approach real-world situations. An interesting feature of all random processes is that they are not totally unpredictable. Only the individual outcome is known. The long-run expectation falls within predictable patterns, such as those explored by us. A minimum of 100 trials are usually needed before a recognizable, valid pattern is obtained.

Practical Steps of Simulation

Having covered the principles and basic nature of simulation, together with illustrative examples, we are ready to consider the practical steps of a simulation study.

Step 1: *Build a Simple Model.* The occasion for construction of a model is the need to investigate and solve a decision problem. Little will be understood until the problem is expressed in terms of the major, crucial factors and variables and their approximate relationships.

Step 2: *Verify Adequacy.* Verify, through checks with persons concerned, that the variables utilized, the ranges of factors considered, and the approximate relations stated have an adequate correspondence (though simplified) to real-world situations.

Step 3: *Make a Diagram.* A flow chart of the major simulation sequences permits the gaining of a good view of the approach to be taken, the amount of detailed programming needed, computer size and memory requirements and the like. At the same time, the absence of important functions or the need to investigate additional relations or to state dependencies and interactions in useful equation form, are highlighted. Corrections, revisions and enlargements can occur accordingly.

Step 4: *Program Development.* Using the diagram and the functions and mathematical model developed, the job of programming for simulation can proceed. Care should be taken to utilize a minimum number of "cases," that is, instances of complete simulation run for particular levels of a variable. Otherwise, the program may quickly become unmanageable. The usual problems of initial program, trial run, errors and omissions, debugging, re-runs, and so on, will consume a great deal of time.

Step 5: *Verify Program.* Using the final runs and the outcomes obtained, check that the results agree with what would be expected at some approximate range from intuition, experience and the like. Do the outcomes look reasonable? Do they reveal internal consistency? Do they mesh with other findings? These are the questions that require close examination. All persons concerned with the simulation study, the gathering of the original data and the use of the end results, should participate in the verification program. It is quite likely that, as a result of these checks, major modifications in the program will be called for. A looping back through Steps 1 through 4 and 5 may then occur, until satisfaction is obtained.

Step 6: *Enrichment Program.* The simplified model and program permit us to evaluate whether or not we are "in the right ball park." Once a determination has been made of the really significant variables, the ranges within which optimal solutions are likely to occur, and the additional factors to be considered, the program should be revised accordingly. We will now add levels of variables, in smaller increments and intervals than before, but within ranges that closely circumscribe the likely solution points. Additional variables can now be added. Further possible alternatives of action can be included in the simulation trials. The entire simulation has thus been enhanced regarding the quality, range and validity of outcomes.

Step 7: *Before-Use Check.* Prior to actual use of the simulation outcomes for the ultimate decision process, verifications similar to Step 5 should occur. Looping back through Steps 6 to 7, or back through earlier stages, may occur. When everyone is satisfied, then and only then, will be sufficient basis for making use of the results of the simulation in the process of making decisions—based on the simulation outcomes and other considerations going beyond the scope of the computer study *per se.*

It should be noted that one cannot at the outset decide that a simulation is needed. It is always possible that an analytical approach is superior. Indeed, ultimately every simulation includes at least certain aspects and facets that involve analytical rather than simulated outcomes.

Simulation versus Analytical Study

Before one knows whether or not a study should be presimulative or analytical in nature, one must have a model. It is from the model, its nature as regards intricacies, number of factors to be considered, levels of significance to be included, and the like, that a judgment is made. The following may be of value in the determination:

1. In general, when a problem can be solved in analytical terms, using equations of relationship, probability calculations and numerical results processing, the analytical approach will be superior. There will be less time consumed in programming and the solution can be obtained more quickly from the computer.
2. Often it is impossible to state the problem in such a way that mathematical analysis techniques can be applied to the solution. A simulation study will then accomplish what cannot be attained in any other way.
3. Sometimes, a valid mathematical formula can be

readily found to tackle the major aspects of a problem. Yet the formula may entail certain features, such as iterative calculation procedures to obtain crucial stepping-stone values of the computations, that it may take hours or more to obtain results—longer than a corresponding simulation.

Further Aspects of Simulation

A certain complementary of methods may be noted. For example, we may use a highly simplified and thus approximate analytical-mathematical formulation to check on a simulation study that has proceeded in greater detail with considerable care. The point is that, despite great pains to use correct values and relationships, errors can always creep into the simulation process. The approximate mathematical formulation then provides check values, useful in addition to the intuitive practical judgment that will be applied in any case.

All of the techniques mentioned as useful adjuncts to model building also apply to simulation: simplicity, enrichment, analogy, looping; as we have noted. The techniques of substituting constants for variables, eliminating redundant or unnecessary steps, permitting linearities to take the place of some curvilinearity, randomizing and strengthening, all apply from modeling in extended form to simulation. A simulation, similar to a model, can be judged for quality on the basis of directness, relatedness, robustness, fertility, flexibility and testability.

For many types of problems, generalized simulation programs have been developed. These can be quickly modified and adjusted to fit a specific set of circumstances. In other instances, a model and the corresponding simulation program will have to be developed *de novo*. Unfortunately, no rules beyond the general ones given here can be provided. The skill and ingenuity of the individuals concerned is the governing factor.

As in previous instances, an awareness of the interlocking

needs of effective simulation brings to the forefront a recognition that close cooperation between programmers, analysts and technical and managerial personnel are essential. Otherwise, loopholes, erroneous or misconceived relationships and other oversights or inadequacies can creep into the simulation, leading to possibly false outcomes.

Computer simulation thus represents a variegated and flexible approach for analysis of decision problems. Participation in such a study will mean that deeper insights and better comprehension of the major problem factors is gained. The interlocking nature of variables can be explored. More informed decisions result.

CHAPTER 26

Implementation of the System

Introduction

Overall performance of management information systems (MIS) depends not only on the equipment, but also on the quality of the program packages, known as the software. These programs make use of basic mill and market data to develop the operating, planning and control information required by management.

It is a matter of painstaking, detailed work to provide computer inputs and outputs in a form that will help the textile company attain its multifold objectives of competitive prices, acceptable quality, and appropriate market share.

A step-by-step sequence of development for the implementation of a program of CIS installation and usage is outlined here.

Overview

Introduction of the information system involves a series of closely interrelated steps. These are:

1. Learning a new vocabulary of technical terms and concepts, so that (a) instructions can be understood and followed, and (b) all persons concerned speak the same language.
2. Study of practical and economic feasibilities of the system.
3. Identification of key systems specifications, required to meet the goals defined as part of the feasibility study.

4. Selection of hardware, that is, the computer equipment, which will permit attainment of the goals of steps 2 and 3.
5. Implementation through application of the appropriate software packages, using computer programs.

Our principal concern presently will be with Step 5, the implementation of the system installation. Software packages are computer programs, usually available in ready-made form, for a company's needs. In the beginning, most applications can be performed with immediately useable programs. In some instances, modifications to meet individual situations, will be called for. As more advanced installations are made, however, it becomes necessary to develop special computer programs tailormade for specific requirements. In all instances the development of appropriate software hinges on the care with which the operating and control needs of managers have been studied, considered and satisfied.

A typical implementation sequence, based on the steps just outlined, appears in Fig. 26-1.

Survey of Current Practices

The business of implementation begins with a survey of current practices and procedures. We ask such questions as:

1. What methods, procedures, practices and policies are currently in use by the company?
2. How well are managers and staff satisfied with current practices?
3. What are the data needs in order to perform optimally?
4. What are typical examples of forms, records and reports containing these essential data?
5. How could these data be supplemented, improved, revised, enlarged to benefit managers and staff in their work?
6. How do people verify the validity of data?

The last point requires further elucidation. Often people

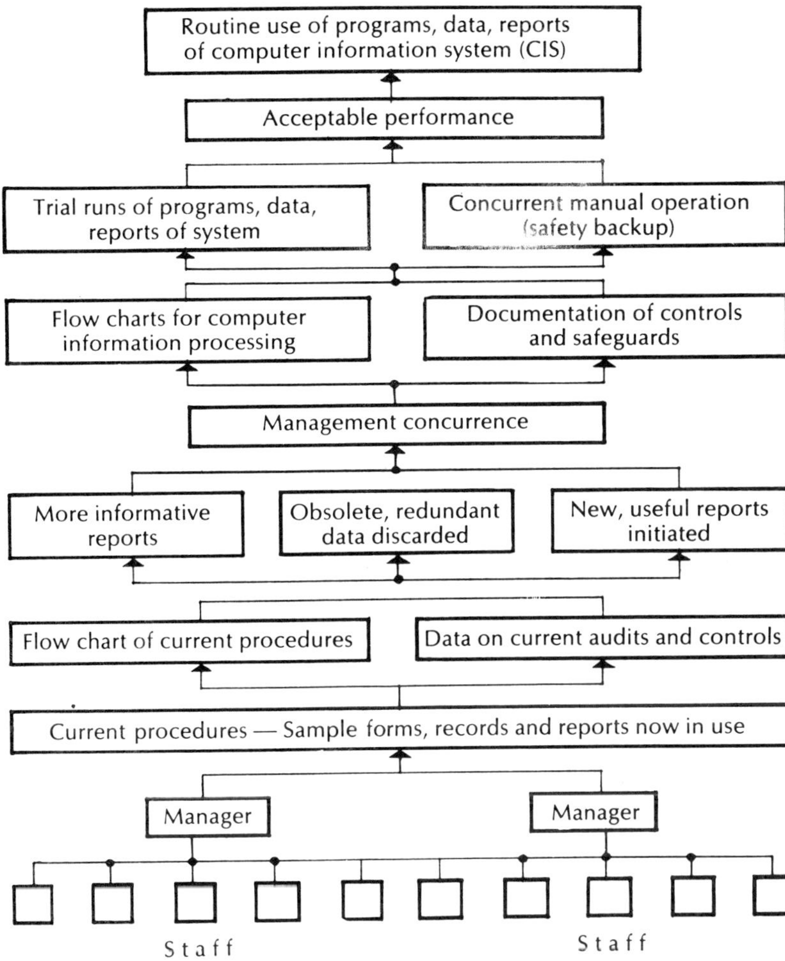

Figure 26 1. Development of program for computer information system (CIS).

have various ways in which to check the accuracy, honesty or basic soundness of data that flow in. It is well to know these checks and to incorporate them in the information system. Then, if an inconsistency or discrepancy beyond allowable error limits occurs, management and staff can be alerted for further checks and investigations.

An example of an inconsistency check is given by the application of a typical production equality:

$$\text{Output from Process A} + \text{Material Added in Process B} - \text{Scrap Loss in Process B} = \text{Input to Process C}$$

The logic of this formula is obvious. The equality sign says in effect that the addition and subtraction on the left-hand side should equal the input on the right-hand side. If there is a major discrepancy (overage or shortage), mill management needs to know this. The computer can be readily programmed to test this equation and issue an "off-standard" report in case of discrepancy that goes beyond allowable limits of normal range of fluctuations in production.

To ascertain the nature of current procedures, it is not enough to talk with supervisors and others, without also observing the users of forms and preparers of records at work in the mill. In this way, aspects of a particular set of procedures not brought out during an interview, will be picked up through actual observation.

Attention must also be paid to the purposes and functions of each paperwork procedure in an interlocking network of operating, planning and control activities. Finally, determinations of frequencies are needed: Hourly, daily, weekly and monthly averages and peak loads. The eventual information system must be geared not just to average requirements but also to accommodate maximum loads.

The information gathered is best recorded and portrayed in flow chart form. Not only does this method provide a ready overview of the procedures network, it will also highlight places where gaps or redundancies occur. Both, primary activities and audits and controls require documentation.

Current Status and Potential Improvements

Reviews and discussions with mill organization executives, managers and staff, aided by flow charts and other documents, will bring the following to light:

1. In many instances, more informative reports would seem helpful. Automated data processing may permit simple, inexpensive development of such improvements.
2. Forms, records and data flows that have become obsolete or are found to be redundant, should be abandoned.
3. Entirely new types of information will be seen to be of considerable value in operating, planning and control activities. Often, these new data can be readily incorporated into the routine outputs of the new information system.
4. Example forms of reports found useful in other companies will often serve as a source of ideas for the present, new installation.

Fig. 26-2 presents a typical report containing information system data outputs.

Full concurrence of executives, managers and staff permit translation of proposals into actual, operating procedures. Just as the pre-MIS procedures were documented in flow-chart form, so should the new practices be put in good form for review and overview.

Usage of the System

An information system consists of a large number of programs that are installed gradually. During this period, there will be trial runs, when both the old and the new system are used simultaneously. When all parties concerned are satisfied that a particular new installation phase is working smoothly, providing the information required in accurate and timely manner, then the manual "backup" may be dropped.

Since major overhauls are costly, the system installed will usually have to serve for a good many years. Nevertheless, there will be ongoing revisions, expansions, innovation and improvements. Control over such activities is maintained through periodic reviews of cost/benefit ratios, computer utilization studies, and information effectiveness analyses.

ACCOUNT PROFITABILITY ANALYSIS

Customer ______________ Season ________ Year ________

	Current Season	Projection for Year	Last Year	Two Years Ago
Gross Sales, $	____	____	____	____
Less Discounts	____	____	____	____
Returns & Allowances	____	____	____	____
Cancellations	____	____	____	____
Net Sales, $	____	____	____	____
Target Sales, $	____	____	____	____
Target as % of Actual Sales	____	____	____	____
Net Pretax Income, $	____	____	____	____
Net Pretax Income as % of Sales	____	____	____	____
Shipping Performance				
Days late	____	____	____	____
Amounts shipped late, $	____	____	____	____
Sales Performance				
Number of salesmen's visits	____	____	____	____
Cost of visits, $	____	____	____	____
Cost as a % of sales	____	____	____	____

Note: Projection equals current year's sales to end of current season, multiplied by a factor based on experience ratios of prior two years.

Figure 26-2. Typical sales building aid obtained from computer information system (CIS).

CHAPTER 27

Motivation to Promote Systems Use

Introduction

From an analysis of numerous information systems installation, the role of motivational factors, combined with technical competence, in the attainment of maximum effectiveness has become widely recognized.

The key to motivation is personal involvement of all of the participants in the eventual system. This includes mill organization executives, management, staff and all users. They all should participate actively in the decisions that lead to installation and use of the information system.

Personal involvement ensures interest, enthusiasm and cooperation in all those activities that represent impelmentation of the change from manual to automated processes.

General

Some highly successful installations of CIS can be found in a large variety of mill production and marketing and sales organizations. When one talks with the people involved, the comment most frequently heard is: "How did we ever get along without our information system?"

Yet, when the history of the program as a whole is reviewed, one usually finds that it was hard to sell. Unlike a new type of production equipment, whose performance can be demonstrated in immediately recognized terms, such as productivity, quality and costs, a computer installation presents many intangibles: Anticipated benefits must be projected on the basis of estimates, case histories of past successes may be only partly relevant and examples involve some conjecture. In the end, the degree of success attained is an individualistic matter, directly related to the amount of personal attention, intensity of motivation and spirit of cooperation and team work that prevail in the mill organization.

Features of Successful Installations

The principal features of information systems that are of real service to a mill organization are:

1. The drudgery of dull routines and repetitive clerical work will have been replaced by automated data processing. People have been freed for more creative tasks having to do with improved customer service and sales, enhanced productivity and better planning and scheduling of operations.
2. The information provided by the system will be timely and will meet the needs of mill management, marketing management, and staff.
3. There will be sufficient flexibility in the system to permit (a) continued adjustments and modifications in accordance with changing marketing, production and financial needs, and (b) improvements and progress in the quality, timeliness and usefulness of the computer services provided.

A high degree of mutual good will among all participants is essential. Data systems and information flows involve intricate networks. Everyone is concerned. Each individual must provide and receive the proper details for his work as part of the overall pattern of administrative, staff, clerical and operating functions of the organization.

Integration of Activities

The crucial need for proper coordination within the information network is brought out in Fig. 27-1. It shows the detail of actions required when a sales order is received. Questions will

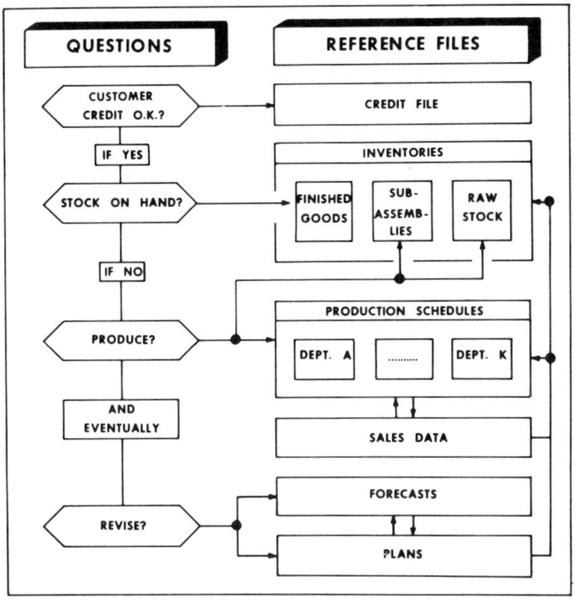

Figure 27-1. Integrated information flow. Typical questions and data needs arising from receipt of a sales order. The success of the system depends on the competence and team work of the designers, who must recognize the many questions, data needs and interrelationships inherent in integrated information flows.

arise regarding customer credit, stock levels, delivery schedules and effects on other activities. Numerous reference files within the information system will need to be consulted.

The interplay between properly stated problem questions and relevant information files will lead to an analysis regarding appropriate decisions. As much of this work as can be done routinely should be automated through computer programming. Numerous individuals will have roles in the installation of these "programmed decision routines." How well are these people motivated towards achievement? How conscious are they of the need for teamwork? The answers to these questions, combined with the technical and professional qualifications of the participants in the CIS, are the determinants of the ultimate degree of success to be expected.

Overcoming Resistance

Much of the initial resistance that may be encountered when ADP is proposed, seems to be rooted in misconceptions about shortcomings of computers. While it is true that earlier computer information systems presented difficulties related to language, timeliness of data, information access, flexibility and costs, these have been largely overcome today. A comparison of old versus new systems appears in Fig. 27-2.

Other areas of concern which must be resolved to everyone's satisfaction are:

1. Data needed for operating decisions must be available in timely and useable form.
2. Controls existing prior to automation should not be weakened by ADP.
3. Such changes of assignments of personnel that may occur should promote, not demote.
4. The organization will become a better place to work in.

The goal is to forestall resistance to major changes, to avoid animosity and to enlist support. There will, of course, be changes in duties of personnel. But with proper training and encouragement, most persons can move on to greater responsibilities. Such work as becomes redundant may be phased out through attrition.

People who have not experienced formal education for a long time may be fearful of training programs. For them, it is important to emphasize that the pace, content and method of development programs will be geared to their particular needs and capabilities.

Proceeding Circumspectly

The receptiveness of managers and staff to changes associated with the CIS can generally be enhanced by practicing good human relations. Among the positive steps to be taken, involving

OLD SYSTEMS		MODERN SYSTEMS
Difficult computer language was involved.	Language	Computer languages are relatively simple. Basics can be learned readily in one or two days.
Long delays, because computer could process only one job or "batch" at a time.	Time	Computer can handle several jobs simultaneously.
Before a specific piece of information in storage could be looked up, the entire data batch would have to be run.	Information Access	Computer goes to specific item directly, gives desired print-out or else a display on cathode ray tube.
Programming and computer use was cumbersome. Only long-planned routines could be run.	Flexibility	New programs, special projects can be run without difficulty. Often, inexpensive small computers are available for specialized tasks. A high degree of versatility is achieved.
High.	Costs	Still high. But the availability of timely, accurate and useful information outweighs costs.

Figure 27-2. Computer capabilities. Modern realities vs. old-time misconceptions.

both human and technical factors, the following are of principal importance:

1. Bring out the ideas of all key people.
   Each manager should help by describing the data he need, by assisting in devising methods for gathering the basic detail, and by helping to develop forms and methods of display of computer outputs. The manager, after all, has the greatest stake in the usefulness of the new installation.
2. Involve the whole staff.
   Staff people, including clerks, often have views of problems at variance with those of their supervisors. Yet, staff and clerical help are the ones who work with detail, making a multitude of small but cumulatively important decisions. Vital facts, essential to the system overall success, will thus be included in the systems design.
3. Question everyone on the types of controls desired. It is not enough to provide for information flows without assuring that checks and controls are included, and that these factors are adequate for managerial and staff requirements.

When truly involved, people will not only provide the essential detail needed to build a good system, but they will also become enghusiastic supporters, helping to bring out numerous improvements throughout the installation and use of the system.

Further Benefits

Some of the benefits of a good system, in addition to those already mentioned, are:

1. Relief of an overburdened staff from routine work.
2. Speedier gathering and processing of data.
3. Fewer time delays.
4. Availability of new types of useful information.

The value of these information benefits is seldom fully understood at the outset. In time, they will be found indispensable.

Caution

One final word of caution: Never develop a new report without first checking it with all those concerned. The new information format is bound to look different from the old one. It may contain more data and be more useful. But do not spring it on anyone without advance warning. Check the new form out first. Explain why you wish to proceed in a certain manner. Show how the new report format will benefit the recipient. A person, once placed in the right frame of mind, will anticipate with pleasure the same report which otherwise he might view with resentment.

CHAPTER 28

Incorporating Management Science in the Information System

Introduction

In this part IV there are essential management aspects of installation of computer-based information systems. Quantitative techniques known as "decision science," "management science," or "operations research" for the analysis and evaluation of complex problems were brought forth in Part III. It is appropriate now that we give attention to the specific manner in which quantitative analysis is incorporated systematically into the overall information serving program of management.

We note three phases of information systems, from initial data compilations to basic types of analyses and, finally, to advanced evaluations. One cannot progress to this third, advanced stage without thoroughly covering the two prior phases, since each stage builds on the information systems developed in the prior stage. We show the interlocking nature of information usage. Also, a time schedule for a normal program is outlined.

General

The history of business-scientific uses of computers is a relatively recent one. Initially, the computer was used merely as a high-speed adding machine, performing the clerical functions of certain accounting and basic management information data including some simple comparisons of performance ratios and related items. But a decade or so ago, management undertook the task of shaping the computer as a tool for providing sophisticated, systems-conscious information designed to serve as a valuable aid in the managerial decision process and based on the principles of management science.

Stages of Development

Regardless of whether we look at the growth of computer usage from the viewpoint of an individual, a progress-minded firm or from the viewpoint of general historical developments, we may identify three stages:

1. The initial stage of management information systems, as outlined in Table 28-1. This stage has three functions: It acquaints and trains people in the design and use of systems that perform various accounting, clerical and basic statistical tasks. It also acquaints people in all levels of management with design features, uses and limitations of systems. We are not talking about "total systems," but rather bits-and-pieces installations such as order processing, accounts payable, payroll, stock levels, waste reports, quality data and so forth. This initial stage also will build files of data for the main data bank.

2. The main stage, summarized in Table 28-2. This utilization leads to the preparation of sales analyses, forecasts, materials explosions, production scheduling, cost control, cash flow, profitability studies, and related endeavors. The source-data developed as a result of the initial stage of automated data processing are combined in various ways to yield these relatively sophisticated analyses.

3. The advanced stage (Table 28-3), reached only in a relatively small number of business organizations. This represents full-scale management science studies to yield information that has a direct bearing on complex decision problems. Some examples are: Product diversification analysis to arrive at an optimal merchandising and promotion program; production smoothing in order to permit proper planning and scheduling of operations;

TABLE 28-1

INITIAL STAGE MANAGEMENT INFORMATION SYSTEMS*		
No.	**Activity**	**Types of Data Built Up as Part of the Activity**
1.1	Order Processing	Volume and dollar amount of sales by season, product line, customer type, geographic area and other breakdowns of interest.
1.2	Invoicing	Weekly volume of shipments, data for use in transportation and distribution planning.
1.3	Accounts receivable	Various types of data for subsequent accounting control.
1.4	Accounts payable	Various types of data for subsequent accounting control.
1.5	Purchasing	Costs of raw materials, components, supplies, repair parts and contracted services.
1.6	Payroll	Labor costs by departments and processes. Piece rates. Cost allocation to products.
1.7	Inventory	Inventory of raw materials, components, supplies, in-process and finished goods.
1.8	Production	Output by departments and products.
1.9	Waste	Costs of repair, rework, seconds, "thirds" and scrap by departments and products.
1.10	Quality	Inspection costs, rejections, customer complaints, costs of field service, returns and allowances.

*These early applications have two purposes: (1) to acquaint and train people in the design and use of systems and (2) to build up files of data that will go into the data banks for the main system.

Source: Enrick, N. L. Management Information Systems course notes, 1977.

TABLE 28-2

MAIN STAGE OF MANAGEMENT INFORMATION SYSTEMS

No.	Activity	Information Provided by the Activity	Source Data*
2.1	Sales analysis	Comparison of this year against last year by months and quarters. Contrasting of actual sales to quotas and forecasts by product lines and sales branches.	1.1
2.2	Sales forecasts	Forecast of anticipated sales, short-term and long-term, based on past experience and calculations, analyses and estimates for future expectations. Forecasting includes periodic revision of anticipations based on actual sales trends and analyses.	1.1, 1.2, 2.2
2.3	Materials explosion	Sales forecasts by products are "exploded" into the raw materials, components, parts, sub-assemblies, supplies, etc., needed to produce the items and have needed inventories ready.	1.1, 1.6 1.8 & 2.1
2.4	Production scheduling	Sales forecasts, materials explosion, production rates, inventories on hand and required, and related detail are used to design production schedules that will assure finished product to be on hand when needed by the customer.	1.1, 1.6, 1.8, 2.1, 2.3
2.5	Cost control	Comparison of actual costs against standards. For example, actual waste (Item 1.9) against allowable waste. Actual production rate (Item 1.8) against production rate implied by payments for piece-work and by piece-rates (Item 1.6). Actual and allowable quality costs (Item 1.10). Costs of carrying inventory (1.7) against inventory needed to support sales (1.1, 2.1, 2.2, 2.3, 2.4)	1.9, 1.8, etc.
2.6	Cash flow	Net inflow or outflow of cash, over a series of past and future expected time periods, based on Items 1.3, 1.4, 2.2, etc.	Many
2.7	Profitability	Advanced studies of costs versus actual and expected value and profitability for all major aspects of operation of the firm: Marketing and promotion vs. returns from this work. Profitability of inventory and warehousing operations. Effectiveness of quality control. Loss rate of customers. Long-term cost-benefit projections.	Many

*In general, the source data shown are merely the basic ones. Additional information must usually be secured. Also, the data bank must be equipped with various programs of requisite degree of refinement to permit precise forecasting (within limits of unavoidable uncertainty), circumspect scheduling, effective marketing and distribution programs, etc. The main stage will thus grow in terms of deliberate, careful steps from the initial stage.

Source: Enrick, N. L. Management Information Systems course notes, 1977.

TABLE 28-3

ADVANCED STAGE OF MANAGEMENT INFORMATION SYSTEMS*

No.	Activity	Information Provided and Methodologies Utilized
3.1	Diversification analysis	Evaluation of present product diversification and determination of optimum degree of such diversification. This analysis weighs the cost of diversification and product proliferation against the needs of customer service and marketing requirements. A technique of linear programming analysis, applied to pertinent data, often reveals the requisite insights.
3.2	Production smoothing	Systematic investigation of a program for smooth production considering output requirements, sales forecasts, manpower requirements and other scheduling factors. Various special analysis models are available, the best known of which is the Holt-Muth-Modigliani model. To smooth production at lowest overall costs, a model geared to the particular firm must be developed, using adaptations of existing models.
3.3	Risk analysis	New projects, involving long-range forward planning and high risks and uncertainties should be analyzed formally. Methods are applied for converting executive opinions, judgement, evaluations and subjective estimates of probabilities (of success or risk of failure) into quantitative terms. Combined with objective data, cost anticipations and profit expectations, the expected value of a long-term project is determined. The data are next refined to discount future earnings on the basis of the value of money (such as 10 per cent per annum). From the specific expected values derived, various proposed projects can be compared and ranked as to respective risks, profitability and overall merit. Choice and selection of those projects that are to be adopted is thus facilitiated. The techniques of "mathematical expectation," decision-free analysis, risk-and-uncertainty evaluation, mathematical probability and futurity discounting are all applicable as aids to the decision process.
3.4	Multi-factor analysis	Many phenomena are the result of many factors. For example, sales volume may be a result of advertising and promotion activity, competitive counter-actions, general economic trends, etc. The technique of multiple regression and multiple correlation analysis permits us to examine data and identify the relative importance and effect of each factor. Future management action can then be based on the relative significance of each factor.

*All of the information in the prior two stages will now be needed. In addition, further data will have to be secured. The methodologies and programs for advanced-stage applications are available, but modifications for each individual firm's special situation are needed. A high degree of sophitication is involved. Stage Two of the Management Information System provides the training, experience and general background that serves as the background and motivation for Stage Three applications.

Source: Enrick, N. L. Management Information Systems course notes, 1977.

risk analysis for new projects; multi-factor analysis for complex, interacting problems to arrive at acceptable solutions.

It is quite apparent, that for advanced management information systems to be performed intelligently and used effectively, close cooperation between management personnel and computer analysts is required. The more knowledgeable management people are, the better will they be able to help the analyst develop mathematical models and relationships that realistically mirror actuality. Similarly, the analyst must have a keen appreciation of the practical facts of business life—the needs, urgencies, behavioral factors and other interlocking factors that may call for a compromise between the theoretical ideal and the practical goal realistically attainable.

Management's Key Role

Some years ago, there was an installation of a relatively advanced management information system in a group of mills of a large firm. When progress in installation and benefits gained was compared among the mills after a two-year effort, it was in those places where management, particularly top management, had taken a real interest in the program, that the greatest progress and the most tangible benefits were identified. Coupled with this interest, we must find a high persistence level. Initial installations almost invariably will be "wrong," because they will have overlooked certain relationships or limitations, no matter how careful management and analyst may have been. Sometimes it takes several revisions until a program works just right and truly gives the decision-serving information and related benefits that are demanded.

Timing Factors

Progress of systems installation and growth in an organization often proceeds too slowly. The schedule in Figure 28-1 represents a rate of progress that seems feasible and attainable in the average organization. It presumes that top management

1. PRELIMINARY DESIGN OF OVERALL SYSTEM — Six Months

a. Define purpose of overall system and identify individual systems needed.
b. Description of methods and procedures to be used for individual systems.
c. Review and revision of basic source documents.
d. Establish ways in which data files, data masters and data banks will be built up from the initial systems.
e. Develop major processing programs.
f. Check on system dependencies, correlation of data sources, elimination of redundancies.

2. HARDWARE AND STAFFING PLANNING — Three Months

a. Review available equipment and decide on that hardware which will best fit the needs of the overall system envisioned.
b. Decide on programming language(s), software needs, consulting services (if any).
c. Determine current and ultimate personnel requirements. Recruit, train, schedule.

3. INITIAL INSTALLATIONS — First Year or Two.

a. Gain experience with installation, use and benefits of systems (staff and management).
b. Computerize simple-to-program applications, usually accounting and clerical functions.
c. From the initial applications, develop data files, masters, data banks.
d. Revise source documents to provide needed input data for master files, proper source-document conversions.
e. Begin to make simple analyses, comparisons, evaluations and compilations.

4. MAIN INSTALLATIONS — Begin in Second Year, Complete by Third or Fourth Year.

a. Build upon and enlarge established data banks.
b. Provide sophisticated comparisons, analyses, evaluations and compilations.
c. Gear (a) and (b) to the managerial scheduling, planning and control functions of the firm.
d. Set-up combined systems, such as sales-inventory-production coordination and plans.

5. ADVANCED INSTALLATIONS — Begin in Third or Fourth Year. Never-Ending Project.

a. Solve complex operations-research type problems involving decision-problems under a variety of alternatives of action and complex, interlocking restrictions on these alternatives.
b. Risk and uncertainty evaluations. General Management Science projects. Correlations.

Figure 28-1. Growth of systems installation and use in an organization. Few textile firms have reached the full state of advanced applications.
Source: Enrick, N. L., Management Information Systems course notes, 1977.

takes a keen interest in Management Information Systems and that there is a degree of astuteness that pushes beyond the initial-installation and main-stage phases towards advanced applications.

Integration of Skills

We have discussed the need for managerial interest and competence of the analyst. In the dynamics of the growth of a system, various factors interact. These involve skills, knowledge and abilities of analysts and programmers, the pertinence of these qualifications to the particular design-phase of the system, and the degree to which there is feedback, willingness to revise, and flexibility of planning in the development of the system. These interlocking relationships are further explored in Figure 28-2. Note particularly how different levels of skills enter as the sophistication of the system increases and the range of applications proliferates. Room is provided not only for growth of soft-ware, but review and revision of hardware availabilities and the training and growth of personnel who must work with the system. The installation of data-processing that leads to truly advanced systems is thus no mean undertaking—rather, it represents a new, formidable challenge to management as regards both its organizing skills and its abilities to acquire the new knowledge that is needed to "make it all work out right."

Quantitatively Oriented Management

Inundated by a constant stream of reports, memos, and data-sheets, many a manager has questioned the desirability to push ahead further in management information systems applications. Nevertheless, a judiciously designed approach can simplify his job, using the exception principle of management.

As a result, an advanced system may push fewer, not more, pages of data-sheets on the manager's desk. But what data are submitted will have been analyzed to provide the essentials

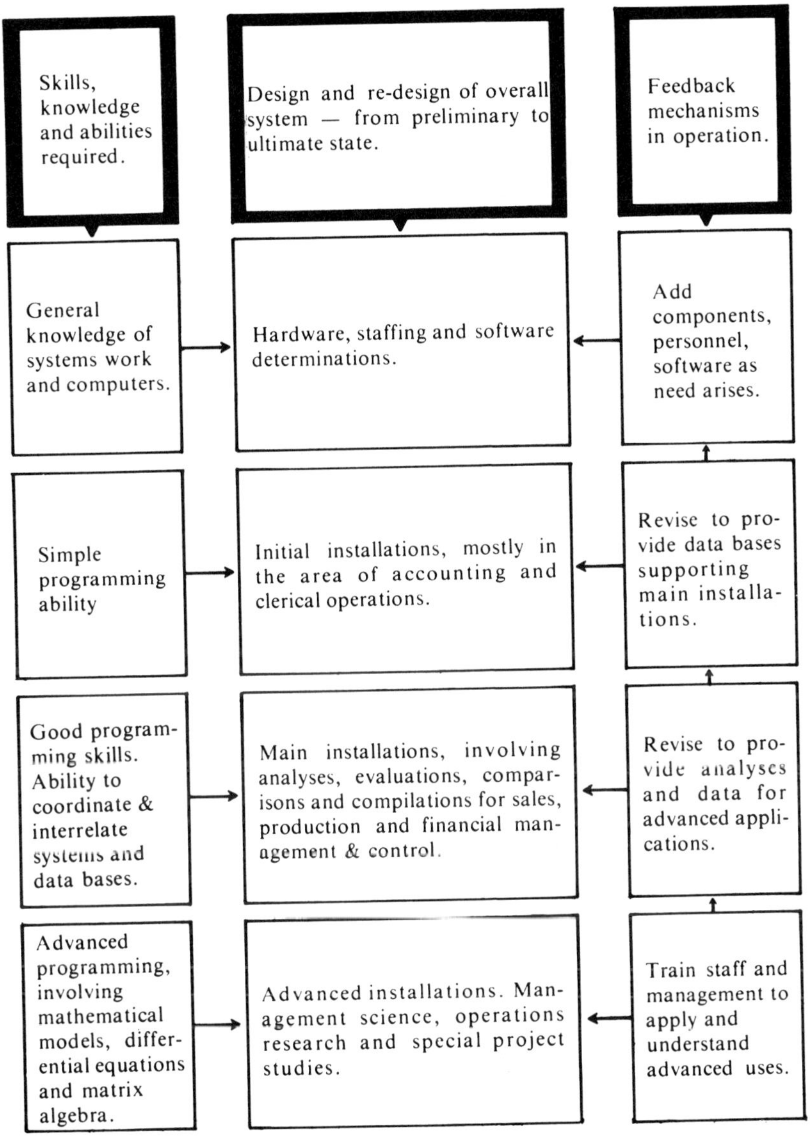

Figure 28-2. Dynamics of systems growth. Versatile skills, effective feedback, and flexible planning are required.

Source: Enrick, N. L., Management Information Systems course notes, 1977.

needed for decision making, and they will certainly highlight only those areas in which "out-of-control" danger situations call for the manager's personal attention.

At all times, the system should serve to lighten, not burden the manager's routine job. The time so gained will now be spent on those activities that no clerk and no computer can handle, activities which call for creative and more interesting use of the executive's talents.

An Application

For an application of operations research methodology in management information systems, we will refer to a product-mix problem example. The three fabric styles, Models X, Y, and Z of the data in Table 28-4, represent a simplification of the 50 styles actually produced by the firm. We show the 1000-yard unit profit, the production rates, and the productive capacity.

Next, utilizing linear programming methodology, an optimum solution is found. It serves in guiding the allocation of priorities of promotional funds for each product. Nevertheless, in practice it is unlikely that orders will come in proportion to promotional effort. Some leeway is needed therefore. The control limits (shown in Line d) provide acceptable deviations, so long as management is willing to be satisfied with an end-result at 95 percent of the ideal, optimal product obtainable. When these control limits are exceeded, the working of the aforementioned exception principle brings management's personal attention into the picture. Among the actions that might be taken would be (1) to check and recheck on the effectiveness of promotion, (2) to permit more than the 5 percent deviation (95 percent ideal limit) previously sought as tolerable, or (3) to make other arrangements. We are not concerned here with the type of action chosen. What is important is that the operations research application of linear programming, combined with the control limits and the exception principle, helps management in (1) achieving a practical, near-optimum result, (2) not being bothered while sales progress is on schedule, and (3) automatically being notified when personal attention and reconsideration are required, leading to new decisions.

TABLE 28-4

LINEAR PROGRAMMING WITH CONTROL LIMITS
Product Mix Example with Optimum Profit Solution and Control Limits

Problem data	Model "X"	Model "Y"	Model "Z"	Productive Capacity hr/wk	Total
a. Profit, in $/unit	90	60	50		
b. Production rate, hours per week					
Spinning	6	18	30	600	
Weaving	8	28	4	400	
Finishing	12	6	4	200	
Problem solution					
c. Optimum quantity for each model, units per week	7.4	10.4	12.3		
d. Control limits, criteria, units per week*	1.5	0.4	1.7		
e. Profit ideally attainable = a x c, in $	666	624	615		1905**
f. Production hours used per week (b x c)					
Spinning	44	187	369		600
Weaving	59	291	49		399***
Finishing	89	62	49		200

*These criteria have been set at 95% of ideal, thus resulting in an expected profit of 0.95 x $1905 = $1810.

**Total is found by adding the profits for each model. No other quantities of output for the three models under consideration can yield a higher profit.

***Because of rounding, this total of 399 does not fully equal the 400 hr. in **b** above. Although for all processes production hours used equal capacity available, this is not necessarily a requirement for optimum profit.

Source: Adapted from N. L. Enrick "Management Planning," ©1967, McGraw-Hill.

The Importance of an Integrated Approach

Earlier, we discussed the three stages of a management information system. One generally must be in stage 1 before one can enter successfully into 2, and similarly for the sequence from stage 2 to 3.

Situations continue to exist where stage 3 work is done with stage 1 programs just beginning. For example, a mill may institute linear programming analysis to coordinate certain phases of sales and production, or to check on desirable levels of product diversification, or to optimize product mix. Such applications then will be isolated, fragmented instances of advanced installations. But they run the risk of sub-optimization: optimizing a component of the total system to the detriment of other factors in the complex as a whole. The development of preliminary, generalized models of the overall systems complex will often go a long way toward reducing the pitfalls in sub-optimization. In general, experience shows that there can be little substitute for the integrated approach that considers all facets of a given and planned future system, as regards both operating procedures and data-flows.

Concluding Observation

From the discussion, it will become apparent why there are few full-fledged, viable, and really profitable advanced programs known. The work involved is sizeable; it requires detailed attention and devotion on the part of many people; and to phase-in, properly develop, and integrate the interdependent stages is not a matter of days and weeks, but of months and years. The payoff from a properly working management information system, as the sparse, available results gathered thus far show, seem worth the effort—as regards both managerial time saving, profitability, and efficiency of operations.

CHAPTER 29

High Speed Innovation

Our age of accelerated technologies has given rise to the cascading of new manufacturing processes and products, and a new kind of staff-management person: the change-seeker. His task is to shorten the conception-to-consumer cycle of new products. Markets for new products tend to be briefer, hence the incentive to get there "first with the mostest." All industries, including textiles, are involved.

The Nature of Accelerated Innovation

We will present a few examples.

1. A fiber producer has engineered a new acrylic. Usually it will take four to five years of research and development, product design, equipment acquisition, marketing analysis, and the like, until the product reaches the consumer. He actually accomplishes all this in less than two years through accelerated planning (See Fig. 29-1).
2. Dissatisfied with the lack of uniformity of available commercial finishes, a textile fabric producers' group is building their own facility. While building is in process, the research laboratory is still working out final details of chemical formulations to obtain a high-quality finish with novel characteristics.
3. A knitter has developed a new line of outerwear of ultramodern styling. While final patterns and color combinations are being evolved, customer lists from the Management Information System are used to solicit opinion-

feedback and also create a favorable attitude among prospective buyers. Time study and industrial engineering are setting up work assignments, materials handling equipment, and layouts of facilities.

Drawbacks

Viewed from traditional precepts of finance, accelerated innovation seems wasteful. Among typical critiques, the following deserve response:

1. How can one justify scrapping of good machinery?

It is hoped that the new products, involving superior appeal and lowered costs, will offset the equipment losses in a relatively short time. A general criterion is that the payback period must not exceed two-and-one-half years.

2. Is it not risky to buy new equipment when product marketing has not been finalized?

Risky, yes. But the new product should have been evaluated through test marketing in three selected, separated areas. The risks are necessary if the firm is to beat its competition to the market place.

3. Isn't all that quick change-over nerve wracking and detrimental to morale?

On the contrary, it is stimulating, by creating sufficient challenge and a sense of quick process. Absence of accelerated innovation produces a feeling of dullness and frustration.

4. How can one set up production lines when the laboratory is still fiddling with final decisions?

The principal features of the product are established. The production line is set-up with allowances for a certain amount of revision, adjustment and redesign. The slight extra effort is justified by the time compression in getting the product to market.

Accelerated Program Flow Chart

The way in which acceleration is planned is illustrated in the flow diagram of Fig. 29-1.

1. During the advanced stages of research and development,

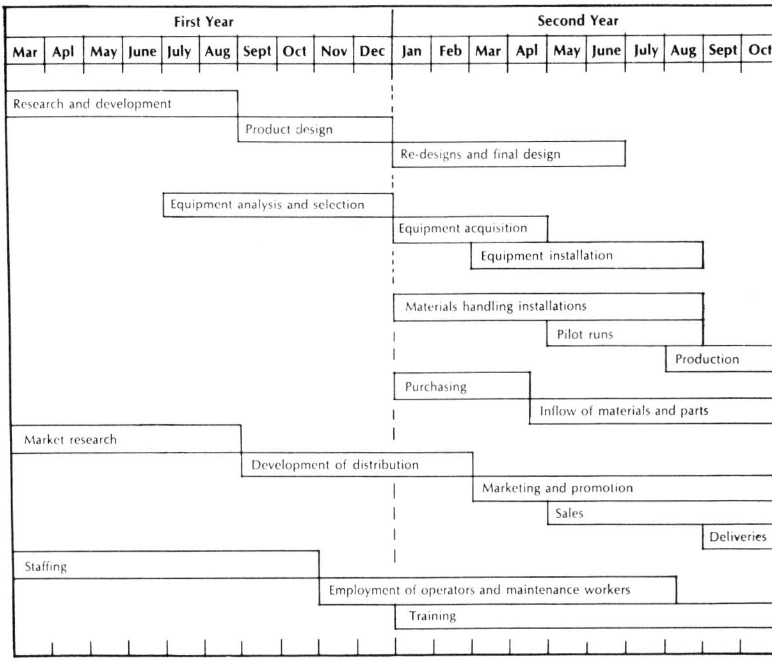

Figure 29-1. Accelerated planning flow diagram. The broad planning of major phases of the quick-react program are supplemented by numerous subsidiary networks of flow charts. Both the over-all and the detailed schedules will require numerous revisions and refinements as work progresses.

and during product design, enough information becomes available to analyze machinery and equipment needs and select the most suitable equipment needs.

2. During product redesign and finalization, incoming equipment can be installed, together with facilities for materials handling.
3. In the later stages of design and equipment installation, pilot runs are planned, to begin as soon as possible, and leading to early regular production.
4. While this work is in process, market research, development of distribution channels, promotion and publicity, as well as sales, all proceed on a synchronized schedule.
5. Staffing, employment and employee training are similarly scheduled. All these synchronous, simultaneous and co-

ordinated activites comprise the system that produces quick-react results.

Maintenance of the accelerated planning flow charts is generally the task of industrial engineering, with the computer information compilations providing the data on expenditures for the project.

Task Management

High-speed innovation involves a number of distinct tasks calling for coordinated management, as shown in Fig. 29-2.

It is not usually appropriate to have full-time managers in these activities. Instead, the managers of various departments, such as Industrial Engineering (whose scope may have been enlarged to encompass Industrial and Management Engineering), Research and Development, General and Production Engineering, Finance, Marketing and Sales are asked to arrange their work in such ways as to free themselves for the coordinating Committee's work.

A finely knit network of communications and coordinated activities is called for among the four major phases shown in the organization of Fig. 29-2. For example, Research and Development jointly with Production and General Engineering are charged with the setting of realistic specifications and tolerances for materials and components. At first these standards apply to semi-final products and for pilot runs, but gradually there must be revisions and adjustments suitable to serve a final product to be shipped to the customer.

Close liaison is needed with Industrial and Management Engineering, which performs the tests and analyzes the quality and reliability of materials, parts and final product. Marketing will have to evaluate whether certain levels of quality and reliability of product are consistent with sales and other needs. For example, a bulky feel may be preferred by the customer, but excessive bulkiness of a textile product may increase flammability, and a compromise must be reached.

Cost benefits analyses seek to balance the cost of quality

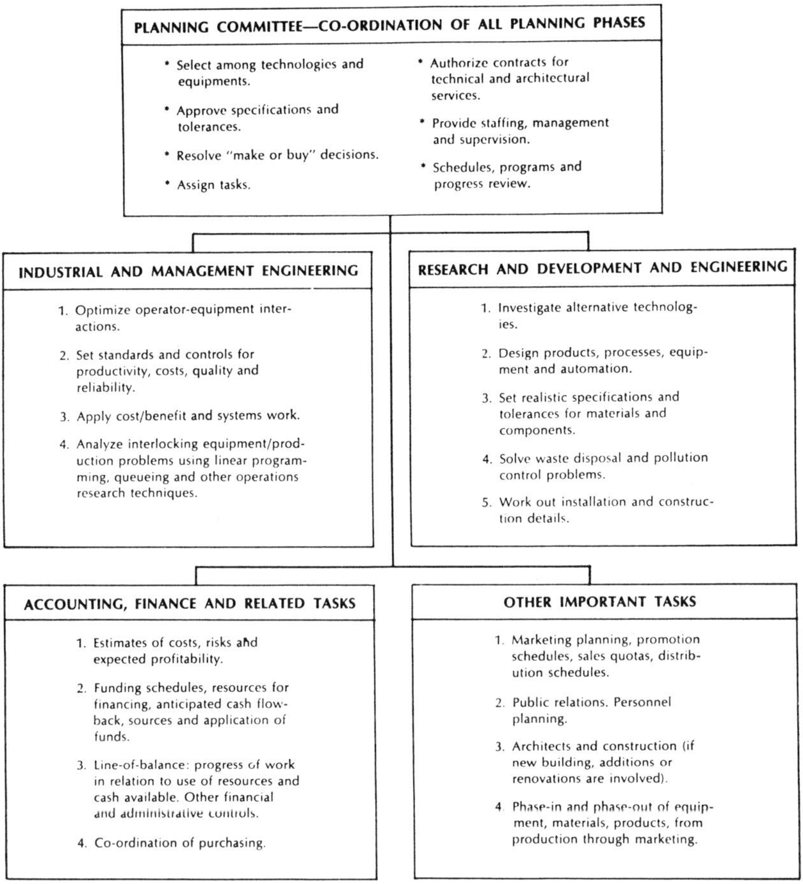

Figure 29-2. Planning committee, organization and functions in implementing innovation tasks.

against the need for precision in processing and for performance in actual use. A series of individual task managers serve the purpose of ironing out subsidiary problems, such as the meshing of spinning and weaving requirements with raw stock availabilities or the gearing together of various time tables.

Coordination problems cannot be minimized. For example, changes in spinning or weaving to overcome productivity problems may, unless checked, create new problems such as in customer acceptance, cost-price relations, or over-all balance of production. As another illustration, a decision to buy yarns (instead of making them on our own spinning equipment) may solve a spinning bottleneck, but will create uncertainty as to whether the vendor can maintain the close control of quality needed and whether he is able to meet promised delivery dates.

Learning Effects

Accelerated planning calls for commitment and intense effort by all participants. Frustrations, lost motions, false starts and backtracking are common—initially. Eventually, with experience, a sense of confidence and accomplishment emerges, as the participants become accustomed to working with quick-react techniques. Under the well-known learning effect, as people gain skill with the new approach, they increase their effectiveness in accomplishing successful high-speed innovations. Most importantly, creativity of the individuals in coming up with useful new ideas and in achieving goals—from product conception to consumer—is enhanced.

APPENDIX A

Methods of Computing Standard Deviations and Variation Coefficients

Introduction

In evaluating observations and measurements, the industrial engineer is not only required to determine the average of the results, but also an indication of variation. The most widely used and accepted measure of variation is the standard deviation, denoted by σ. When the standard deviation is expressed as a per cent of the average of the results, the coefficient of variation, V, is obtained (also denoted by C.V.). The various uses of these two measures in industrial engineering work have been shown throughout this manual. There are, however, several methods of computing standard deviations, depending upon the purpose involved, and these methods are described herein. For each method the additional step giving the variation coefficient is also given.

Methods Shown

The methods presented are given in the order of their relative development. The basic method is shown in Table A-1. Now, when it is desired to estimate the variation coefficient of a production process or a shipment or lot *from a sample*, the basic method shown in Table A-1 will usually slightly underestimate the true variation coefficient. In order to correct for this bias, the quantity N should be replaced by N-1 (N is the sample size and N-1 is known as the *Degrees of Freedom*). The method is illustrated in Table A-2. It will be noted that as N becomes larger, the

TABLE A-1
Basic Method of Computing Variation Coefficient
(Example of Tensile Strength Determinations)

Tensile strength, in pounds, X	Average, $\bar{X}$	Deviation from average, $X - \bar{X}$	Deviation from average squared, $(X - \bar{X})^2$
64	62	+2	4
60	62	−2	4
62	62	0	0
66	62	+4	16
58	62	−4	16
310		0	40

Average, $\bar{X} = \Sigma X/N = 310/5 = 62$, where N equals the sample size
Variance, $\sigma^2 = \Sigma(X - \bar{X})^2/N = 40/5 = 8$
Standard deviation, $\sigma = \sqrt{\sigma^2} = \sqrt{8} = 2.828$
Variation coefficient, $\hat{V}\% = (\sigma/\bar{X})\ 100 = (2.82/62)\ 100 = 4.56$

TABLE A-2
Estimating Coefficient of Variation of a Lot from a Sample
(Example of Tensile Strength Determinations)

Tensile strength, in pounds, X	Average, $\bar{X}$	Deviation from average, $X - \bar{X}$	Deviation from average, squared, $(X - \bar{X})^2$
64	62	+2	4
60	62	−2	4
62	62	0	0
66	62	+4	16
58	62	−4	16
310		0	40

Average, $\bar{X} = \Sigma X/N = 310/5 = 62$
Variance, $\hat{\sigma}^2 = \Sigma(X - \bar{X})^2/N - 1 = 40/4 = 10$
Standard deviation, $\hat{\sigma} = \sqrt{\hat{\sigma}^2} = \sqrt{10} = 3.162$
Variation coefficient, $\hat{V}\% = (\sigma/\bar{X})100 = (3.162/62)100 = 5.10$
The caret, ^, denotes that this is an estimate of the lot σ obtained from a sample.

TABLE A-3
Estimating Coefficient of Variation of a Production Lot from a Sample, Short Cut Method of Computation (Example of Tensile Strength Determinations)

Tensile strength, in pounds, X	X^2
64	4,096
60	3,600
62	3,844
66	4,356
58	3,364
310	19,260

$(\Sigma X)^2 = 310^2 = 96{,}100$
Average, $\bar{X} = 62$
$N = 5;\ N - 1 = 4;\ N(N-1) = 20$
Variance, $\hat{\sigma}^2 = [\Sigma X^2/(N-1)] - \{(\Sigma X)^2/[N(N-1)]\}$
$= (19{,}260/4) - (96{,}100/20) = 10$
Standard deviation, $\hat{\sigma} = \sqrt{\hat{\sigma}^2} = \sqrt{10} = 3.162$
Variation coefficient, $\hat{V}\% = (3.162/62)100 = 5.10$

TABLE A-4
Estimating Coefficient of Variation of a Lot from a Sample, Short Cut Method of Computation for Grouped Data (Example of Product Weight)

Product weight, ounces	Midvalue, ounces	Frequency, f	Deviation,[a] d	fd	fd^2
1.0–1.19	1.15	2	−3	−6	18
1.2–1.29	1.25	2	−2	−4	8
1.3–1.39	1.35	4	−1	−4	4
1.4–1.49	1.45	6	0	0	0
1.5–1.59	1.55	3	+1	+3	3
1.6–1.69	1.65	1	+2	+2	4
1.7–1.80	1.75	2	+3	+6	18
		20		−3	55

Average, $\bar{X} = A + \Sigma(fd)(i)/N = 1.45 + [(-3)(0.1)/20] = 1.435$
Variance, $\hat{\sigma}^2 = \{[\Sigma fd^2/(N-1)] - (\Sigma fd)^2/[N(N-1)]\}i^2$
$= \{(55/19) - (-3)^2/[(20)(19)]\}0.1^2 = 0.0287$
Standard deviation $= \sqrt{\hat{\sigma}^2} = \sqrt{0.0287} = 0.169$
Variation coefficient, $\hat{V}\% = (\hat{\sigma}/\bar{X})100 = (0.169/1.435)100 = 11.8$

[a] Deviations are measured in full class interval units, from a tentatively assumed average, A. In this example, A was taken at 1.45. The class interval units, i, are the differences between successive midvalues, or 0.1 oz. (for example, 1.25 − 1.15 is 0.1).

TABLE A-5
Breaking Strength of Fabric Samples

Lab. no.	Sample A Observations	Sample A Sum	Sample B Observations	Sample B Sum	Sample C Observations	Sample C Sum	Total of the sums	Average observation
1	42.2 + 41.7 =	83.9	42.2 + 41.5 =	83.7	43.9 + 41.3 =	85.2	252.8	42.13
2	37.8 + 37.5 =	75.3	36.7 + 36.2 =	72.9	36.5 + 36.6 =	73.1	221.3	36.88
3	33.4 + 33.8 =	67.2	34.2 + 33.7 =	67.9	35.8 + 35.4 =	71.2	206.3	34.38
4	40.5 + 42.3 =	82.8	40.1 + 42.1 =	82.2	39.2 + 40.3 =	79.5	244.5	40.75
Total		309.2		306.7		309.0	924.9	154.14
Average		38.65		38.34		38.62		38.54

Grand total = 924.9
Grand average = 924.9/24 = 38.54

TABLE A-6
Analysis of Variance Calculations for Breaking Strength of Samples of Fabric from Various Laboratories

	Sources of variation			
	1. Between labs.	2. Between samples within labs.	3. Experimental error	4. Total
a. Add squared totals	252.8^2 221.3^2 206.3^2 $+\ 244.5^2$ ——— 215,221	83.9^2 75.3^2 $+\ 67.2^2$ ——— etc. for all 12 totals		42.2^2 37.8^2 $+\ 33.4^2$ ——— etc. for all 24 totals
b. Sum of entries in *a*	215,221	71,761		35,889
c. Number of tests for each total in *a*	6	2		1
d. b/c	35,870	35,880		35,889
e. Correction factor[a]	35,643	35,643		35,643
f. Formula for sum of squares	$d_1 - e$	$d_2 - d_1$	$g_4 - g_1 - g_2$	$d_4 - e$
g. Sum of squares	227	10	9	246
h. Degrees of freedom	$4 - 1 = 3$	$4(3 - 1) = 8$	12	$24 - 1 = 23$
i. Mean square, g/h	75.67	1.25	0.75	
j. F-ratio, i/i_3	100.89	1.67		
k. Significance level[b] (from a table of F-ratios)	99.9%	Nil		
l. Formula for variance	$(i_1 - i_2)/6$	$(i_2 - i_3)/2$	i_3	$m_1 + m_2 + m_3$
m. Variance	12.40	0.25	0.75	13.40

[a] (Grand total)^2/no. of tests = $924.9^2/24$ = 35,643.

[b] Tables of F-ratios and corresponding significance levels are widely published in statistical texts, e.g., A. H. Duncan, *Quality Control and Industrial Statistics*, R. D. Irwin, Homewood, Ill.

difference between this quantity and N-1 becomes negligible, so that many texts use N in place of N-1 wherever the sample size exceeds 25.

When a desk calculator is available, and also when using electronic computers, it is simpler to use the short cut method in Table A-3 which yields answers identical to those found by the method of Table A-2. As a further step, when the data are grouped in the form of a frequency distribution (as for example in Table A-4) the computations proceed as shown in this table.

In standardization work it is often necessary to evaluate the relative magnitude of various sources of variation. When this is required, it is usually desirable to compute the variances, representing the squared standard deviations. As an illustration, the data in Table A-5 represent breaking strength determinations on three samples of fabric, two tests or observations per fabric, in each of four laboratories. In order to evaluate the effect of differences between laboratories and between samples within each laboratory, variances are calculated as shown in Table A-6. The last line provides the results, showing that there is a relatively large difference between laboratories, whereas the difference between samples and the unexplained effects or "experimental error" are relatively small. Thus, it is evident that a program is required to standardize the laboratory testing procedures, so as to obtain comparable test results in determining fabric quality. An analysis of the sources of variation indicates where additional work is needed to obtain standardization of operations, observations, and measurements.

APPENDIX B

Areas Under the Normal Curve

TABLE B-1
Simplified Table

Range of std. dev.[a]	Items falling within this range, %[b]	Items falling outside this range, %[b]
0.1	4.0	46.0
0.2	8.0	42.0
0.3	11.8	38.2
0.4	15.5	34.5
0.5	19.1	30.9
0.6	22.6	27.4
0.7	25.8	24.2
0.8	28.8	21.2
0.9	31.6	18.4
1.0	34.1	15.9
1.1	36.4	13.6
1.2	38.5	11.5
1.3	40.3	9.7
1.4	41.9	8.1
1.5	43.3	6.7
1.6	44.5	5.5
1.8	46.4	3.6
1.9	47.1	2.9
2.0	47.7	2.3
2.3	48.9	1.1
2.6	49.5	0.5
3.0	49.9	0.1

[a] Range to either side of the distribution average.

[b] In place of the term "items" the terms units, articles, values, observations, or measurements may be substituted, depending upon the particular use of the table.

Example 1. To find the per cent of items falling within a range of 1 standard deviation on either side of the distribution average, read 34.1 per cent in the second column opposite 1.0 in the first column. Since under the normal curve 50 per cent

TABLE B-2
Detailed Table

Range of std. dev.	Items within std. dev. range, %	Range of std. dev.	Items within std. dev. range, %	Range of std. dev.	Items within std. dev. range, %	Range of std. dev.	Items within std. dev. range, %
0.01	0.399	0.42	16.276	0.83	29.673	1.24	39.251
0.02	0.798	0.43	16.640	0.84	29.955	1.25	39.435
0.03	1.197	0.44	17.003	0.85	30.234	1.26	39.617
0.04	1.595	0.45	17.364	0.86	30.511	1.27	39.796
0.05	1.994	0.46	17.724	0.87	30.785	1.28	39.973
0.06	2.392	0.47	18.082	0.88	31.057	1.29	40.147
0.07	2.790	0.48	18.439	0.89	31.327	1.30	40.320
0.08	3.188	0.49	18.793	0.90	31.594	1.31	40.490
0.09	3.586	0.50	19.146	0.91	31.859	1.32	40.658
0.10	3.983	0.51	19.497	0.92	32.121	1.33	40.824
0.11	4.380	0.52	19.847	0.93	32.381	1.34	40.988
0.12	4.776	0.53	20.194	0.94	32.639	1.35	41.149
0.13	5.172	0.54	20.540	0.95	32.894	1.36	41.309
0.14	5.567	0.55	20.884	0.96	33.147	1.37	41.466
0.15	5.962	0.56	21.226	0.97	33.398	1.38	41.621
0.16	6.356	0.57	21.566	0.98	33.646	1.39	41.774
0.17	6.749	0.58	21.904	0.99	33.891	1.40	41.924
0.18	7.142	0.59	22.240	1.00	34.134	1.41	42.073
0.19	7.535	0.60	22.575	1.01	34.375	1.42	42.220
0.20	7.926	0.61	22.907	1.02	34.614	1.43	42.364
0.21	8.317	0.62	23.237	1.03	34.850	1.44	42.507
0.22	8.706	0.63	23.565	1.04	35.083	1.45	42.647
0.23	9.095	0.64	23.891	1.05	35.314	1.46	42.786
0.24	9.483	0.65	24.215	1.06	35.543	1.47	42.922
0.25	9.871	0.66	24.537	1.07	35.769	1.48	43.056
0.26	10.257	0.67	24.857	1.08	35.993	1.49	43.189
0.27	10.642	0.68	25.175	1.09	36.214	1.50	43.319
0.28	11.026	0.69	25.490	1.10	36.433	1.51	43.448
0.29	11.409	0.70	25.804	1.11	36.650	1.52	43.574
0.30	11.791	0.71	26.115	1.12	36.864	1.53	43.699
0.31	12.172	0.72	26.424	1.13	37.076	1.54	43.822
0.32	12.552	0.73	26.730	1.14	37.286	1.55	43.943
0.33	12.930	0.74	27.035	1.15	37.493	1.56	44.062
0.34	13.307	0.75	27.337	1.16	37.698	1.57	44.179
0.35	13.683	0.76	27.637	1.17	37.900	1.58	44.295
0.36	14.058	0.77	27.935	1.18	38.100	1.59	44.408
0.37	14.431	0.78	28.230	1.19	38.298	1.60	44.520
0.38	14.803	0.79	28.524	1.20	38.493	1.61	44.630
0.39	15.173	0.80	28.814	1.21	38.686	1.62	44.738
0.40	15.542	0.81	29.103	1.22	38.877	1.63	44.845
0.41	15.910	0.82	29.389	1.23	39.065	1.64	44.950

(continued)

TABLE B-2 *(continued)*

Range of std. dev.	Items within std. dev. range, %	Range of std. dev.	Items within std. dev. range, %	Range of std. dev.	Items within std. dev. range, %	Range of std. dev.	Items within std. dev. range, %
1.65	45.053	2.06	48.030	2.47	49.324	2.88	49.801
1.66	45.154	2.07	48.077	2.48	49.343	2.89	49.807
1.67	45.254	2.08	48.124	2.49	49.361	2.90	49.813
1.68	45.352	2.09	48.169	2.50	49.379	2.91	49.819
1.69	45.449	2.10	48.214	2.51	49.396	2.92	49.825
1.70	45.543	2.11	48.257	2.52	49.413	2.93	49.831
1.71	45.637	2.12	48.300	2.53	49.430	2.94	49.836
1.72	45.728	2.13	48.341	2.54	49.446	2.95	49.841
1.73	45.818	2.14	48.382	2.55	49.461	2.96	49.846
1.74	45.907	2.15	48.422	2.56	49.477	2.97	49.851
1.75	45.994	2.16	48.461	2.57	49.492	2.98	49.856
1.76	46.080	2.17	48.500	2.58	49.506	2.99	49.861
1.77	46.164	2.18	48.537	2.59	49.520	3.00	49.865
1.78	46.246	2.19	48.574	2.60	49.534	3.01	49.869
1.79	46.327	2.20	48.610	2.61	49.547	3.02	49.874
1.80	46.407	2.21	48.645	2.62	49.560	3.03	49.878
1.81	46.485	2.22	48.679	2.63	49.573	3.04	49.882
1.82	46.562	2.23	48.713	2.64	49.585	3.05	49.886
1.83	46.638	2.24	48.745	2.65	49.598	3.06	49.889
1.84	46.712	2.25	48.778	2.66	49.609	3.07	49.893
1.85	46.784	2.26	48.809	2.67	49.621	3.08	49.897
1.86	46.856	2.27	48.840	2.68	49.632	3.09	49.900
1.87	46.926	2.28	48.870	2.69	49.643	3.10	49.903
1.88	46.995	2.29	48.899	2.70	49.653	3.11	49.906
1.89	47.062	2.30	48.928	2.71	49.664	3.12	49.910
1.90	47.128	2.31	48.956	2.72	49.674	3.13	49.913
1.91	47.193	2.32	48.983	2.73	49.683	3.14	49.916
1.92	47.257	2.33	49.010	2.74	49.693	3.15	49.918
1.93	47.320	2.34	49.036	2.75	49.702	3.16	49.921
1.94	47.381	2.35	49.061	2.76	49.711	3.17	49.924
1.95	47.441	2.36	49.086	2.77	49.720	3.18	49.926
1.96	47.500	2.37	49.111	2.78	49.728	3.19	49.929
1.97	47.558	2.38	49.134	2.79	49.736	3.20	49.931
1.98	47.615	2.39	49.158	2.80	49.744	3.21	49.934
1.99	47.670	2.40	49.180	2.81	49.752	3.22	49.936
2.00	47.725	2.41	49.202	2.82	49.760	3.23	49.938
2.01	47.778	2.42	49.224	2.83	49.767	3.24	49.940
2.02	47.831	2.43	49.245	2.84	49.774	3.25	49.942
2.03	47.882	2.44	49.266	2.85	49.781	3.26	49.944
2.04	47.932	2.45	49.286	2.86	49.788	3.27	49.946
2.05	47.982	2.46	49.305	2.87	49.795	3.28	49.948

TABLE B-2 *(continued)*

Range of std. dev.	Items within std. dev. range, %	Range of std. dev.	Items within std. dev. range, %	Range of std. dev.	Items within std. dev. range, %	Range of std. dev.	Items within std. dev. range, %
3.29	49.950	3.47	49.974	3.65	49.987	3.83	49.994
3.30	49.952	3.48	49.975	3.66	49.987	3.84	49.994
3.31	49.953	3.49	49.976	3.67	49.988	3.85	49.994
3.32	49.955	3.50	49.977	3.68	49.988	3.86	49.994
3.33	49.957	3.51	49.978	3.69	49.989	3.87	49.995
3.34	49.958	3.52	49.978	3.70	49.989	3.88	49.995
3.35	49.960	3.53	49.979	3.71	49.990	3.89	49.995
3.36	49.961	3.54	49.980	3.72	49.990	3.90	49.995
3.37	49.962	3.55	49.981	3.73	49.990	3.91	49.995
3.38	49.964	3.56	49.981	3.74	49.991	3.92	49.996
3.39	49.965	3.57	49.982	3.75	49.991	3.93	49.996
3.40	49.966	3.58	49.983	3.76	49.992	3.94	49.996
3.41	49.968	3.59	49.983	3.77	49.992	3.95	49.996
3.42	49.969	3.60	49.984	3.78	49.992	3.96	49.996
3.43	49.970	3.61	49.985	3.76	49.992	3.97	49.996
3.44	49.971	3.62	49.985	3.80	49.993	3.98	49.997
3.45	49.972	3.63	49.986	3.81	49.993	3.99	49.997
3.46	49.973	3.64	49.986	3.82	49.993	4.00	49.998

of the items lie on either side of the distribution average, 50 − 34.1 or 15.9 per cent of the items fall outside of a range of 1 standard deviation on either side of the distribution average, as shown in the third column.

Example 2. As shown in the second column, 49.5 per cent of all items fall within a range of 2.6 standard deviation on either side of the distribution average and, as shown in the third column, 0.5 per cent fall outside.

APPENDIX C

Tables of Productivity Factors

The productivity factors A for various cycle ratios P and numbers N of machines assigned to an operator, presented here, are an expansion of Table 15-5. They are calculated on the assumption, based on textile industry experience, of a Poisson distribution of arrival rates of stops and a negative exponential distribution of service times. For each value of P and N the corresponding A is listed. Machine efficiency is then found from A/N.

Application of these tables is illustrated in Chapter 15, Machine Interference.

While the tables here should be applicable for the predominant textile industrial engineering needs, those interested will find expanded tables and further examples of applications in a book by O'Connor (1).

Reference

1. O'Connor, Thomas F., *Profitable Productivity*, Manchester, England: Emmot & Co. Ltd., 1966.

TABLE C-1

	Cycle Ratio, P						
N	.010	.012	.014	.016	.018	.020	.022
1	.990	.988	.986	.984	.982	.980	.978
2	1.980	1.976	1.972	1.968	1.964	1.960	1.956
3	2.970	2.964	2.957	2.951	2.945	2.939	2.933
4	3.959	3.951	3.942	3.934	3.925	3.917	3.908
5	4.948	4.938	4.927	4.916	4.905	4.894	4.883
6	5.938	5.924	5.911	5.898	5.884	5.870	5.856
7	6.926	6.911	6.895	6.879	6.862	6.845	6.828
8	7.915	7.897	7.878	7.859	7.839	7.819	7.799
9	8.903	8.882	8.861	8.839	8.816	8.792	8.768
10	9.892	9.868	9.843	9.818	9.791	9.764	9.736
11	10.879	10.853	10.825	10.796	10.766	10.735	10.702
12	11.867	11.837	11.806	11.773	11.739	11.704	11.667
13	12.854	12.821	12.786	12.750	12.712	12.672	12.630
14	13.841	13.805	13.766	13.725	13.683	13.638	13.590
15	14.828	14.788	14.745	14.700	14.653	14.602	14.549
16	15.815	15.771	15.724	15.674	15.621	15.565	15.505
17	16.801	16.753	16.702	16.647	16.588	16.526	16.459
18	17.787	17.735	17.679	17.619	17.554	17.484	17.410
19	18.772	18.716	18.655	18.589	18.518	18.441	18.358
20	19.757	19.696	19.630	19.558	19.480	19.395	19.303
21	20.742	20.677	20.605	20.527	20.441	20.347	20.244
22	21.727	21.656	21.579	21.493	21.399	21.296	21.182
23	22.711	22.635	22.551	22.459	22.356	22.242	22.115
24	23.695	23.613	23.523	23.422	23.310	23.185	23.045
25	24.678	24.591	24.493	24.384	24.262	24.124	23.969
26	25.661	25.568	25.463	25.345	25.211	25.060	24.888
27	26.643	26.544	26.431	26.303	26.158	25.991	25.801
28	27.625	27.519	27.398	27.260	27.101	26.919	26.708
29	28.607	28.494	28.364	28.215	28.042	27.841	27.608
30	29.588	29.467	29.328	29.167	28.978	28.758	28.500
31	30.568	30.440	30.291	30.117	29.912	29.670	29.384
32	31.548	31.412	31.252	31.064	30.841	30.575	30.259
33	32.528	32.383	32.212	32.008	31.765	31.474	31.123
34	33.507	33.353	33.170	32.950	32.685	32.365	31.976
35	34.485	34.322	34.126	33.888	33.600	33.248	32.817
36	35.463	35.289	35.079	34.824	34.509	34.122	33.644
37	36.440	36.256	36.031	35.755	35.412	34.986	34.456
38	37.416	37.221	36.981	36.682	36.309	35.840	35.251
39	38.392	38.185	37.928	37.605	37.198	36.681	36.029
40	39.367	39.147	38.872	38.524	38.079	37.510	36.786
41	40.341	40.108	39.814	39.438	38.952	38.325	37.522
42	41.315	41.068	40.753	40.346	39.816	39.124	38.233
43	42.287	42.026	41.688	41.248	40.669	39.907	38.919
44	43.259	42.982	42.621	42.145	41.511	40.671	39.576
45	44.230	43.936	43.549	43.034	42.341	41.414	40.202
46	45.200	44.888	44.474	43.916	43.158	42.135	40.797
47	46.169	45.838	45.395	44.790	43.960	42.833	41.356
48	47.137	46.786	46.311	45.655	44.746	43.504	41.879
49	48.104	47.732	47.222	46.511	45.515	44.147	42.363
50	49.070	48.675	48.128	47.356	46.265	44.761	42.808

TABLE C-2

	Cycle Ratio, P						
N	.010	.012	.014	.016	.018	.020	.022
51	50.034	49.616	49.029	48.190	46.994	45.342	43.212
52	50.997	50.554	49.923	49.013	47.702	45.889	43.575
53	51.959	51.489	50.811	49.821	48.386	46.401	43.898
54	52.920	52.420	51.692	50.616	49.044	46.876	44.181
55	53.879	53.348	52.566	51.395	49.674	47.313	44.426
56	54.837	54.273	53.431	52.157	50.276	47.710	44.635
57	55.793	55.194	54.288	52.901	50.846	48.069	44.810
58	56.748	56.111	55.134	53.625	51.383	48.389	44.955
59	57.700	57.023	55.971	54.328	51.886	48.671	45.073
60	58.651	57.930	56.796	55.008	52.354	48.917	
61	59.600	58.833	57.610	55.663	52.785	49.127	
62	60.547	59.730	58.410	56.293	53.179	49.306	
63	61.492	60.621	59.196	56.894	53.536	49.455	
64	62.434	61.506	59.967	57.466	53.856	49.578	
65	63.374	62.384	60.721	58.008	54.140		
66	64.311	63.256	61.458	58.517	54.389		
67	65.246	64.119	62.176	58.993	54.605		
68	66.177	64.974	62.874	59.435	54.790		
69	67.106	65.820	63.549	59.842	54.946		
70	68.031	66.657	64.202	60.213	55.076		
71	68.953	67.483	64.830	60.550			
72	69.871	68.298	65.432	60.852			
73	70.785	69.101	66.006	61.120			
74	71.695	69.891	66.552	61.356			
75	72.601	70.668	67.068	61.561			
76	73.501	71.429	67.553	61.737			
77	74.397	72.175	68.005	61.887			
78	75.287	72.904	68.426	62.013			
79	76.172	73.615	68.813				
80	77.051	74.307	69.167				
81	77.923	74.978	69.488				
82	78.787	75.627	69.778				
83	79.645	76.254	70.035				
84	80.494	76.856	70.263				
85	81.335	77.432	70.463				
86	82.167	77.983	70.635				
87	82.989	78.505	70.783				
88	83.801	78.999					
89	84.602	79.463					
90	85.390	79.898					
91	86.166	80.302					
92	86.929	80.675					
93	87.677	81.017					
94	88.410	81.330					
95	89.126	81.612					
96	89.826	81.865					
97	90.507	82.091					
98	91.169	82.290					
99	91.810	82.464					
100	92.430	82.615					

TABLE C-3

	Cycle Ratio, P						
N	.024	.026	.028	.030	.032	.034	.036
1	.977	.975	.973	.971	.969	.967	.965
2	1.952	1.948	1.944	1.940	1.936	1.932	1.928
3	2.926	2.920	2.914	2.908	2.901	2.895	2.889
4	3.900	3.891	3.882	3.873	3.864	3.855	3.846
5	4.871	4.860	4.848	4.836	4.825	4.813	4.801
6	5.842	5.827	5.813	5.798	5.783	5.767	5.752
7	6.811	6.793	6.775	6.756	6.738	6.718	6.699
8	7.778	7.757	7.735	7.712	7.689	7.666	7.642
9	8.744	8.718	8.692	8.665	8.638	8.609	8.581
10	9.707	9.677	9.647	9.615	9.582	9.549	9.514
11	10.669	10.634	10.598	10.561	10.523	10.483	10.441
12	11.629	11.589	11.547	11.503	11.458	11.411	11.362
13	12.586	12.540	12.492	12.441	12.389	12.334	12.276
14	13.540	13.488	13.433	13.375	13.314	13.249	13.182
15	14.492	14.433	14.369	14.303	14.232	14.157	14.079
16	15.441	15.373	15.301	15.225	15.143	15.057	14.965
17	16.387	16.310	16.228	16.140	16.047	15.946	15.840
18	17.329	17.242	17.149	17.049	16.941	16.825	16.701
19	18.267	18.170	18.064	17.949	17.826	17.692	17.548
20	19.202	19.091	18.972	18.841	18.699	18.545	18.378
21	20.131	20.007	19.871	19.723	19.560	19.382	19.188
22	21.056	20.916	20.763	20.593	20.406	20.202	19.978
23	21.975	21.818	21.644	21.451	21.237	21.002	20.743
24	22.887	22.712	22.515	22.295	22.051	21.780	21.481
25	23.794	23.596	23.374	23.124	22.844	22.533	22.189
26	24.692	24.471	24.219	23.935	23.615	23.258	22.863
27	25.583	25.334	25.049	24.726	24.361	23.953	23.500
28	26.464	26.185	25.863	25.496	25.080	24.613	24.097
29	27.337	27.022	26.659	26.241	25.767	25.236	24.649
30	28.198	27.844	27.433	26.960	26.421	25.818	25.154
31	29.046	28.649	28.185	27.649	27.039	26.356	25.610
32	29.882	29.436	28.912	28.305	27.616	26.848	26.016
33	30.702	30.201	29.611	28.926	28.150	27.292	26.370
34	31.506	30.944	30.279	29.508	28.639	27.685	26.673
35	32.292	31.661	30.913	30.049	29.080	28.029	26.928
36	33.058	32.350	31.512	30.547	29.474	28.324	27.137
37	33.802	33.009	32.072	30.999	29.818	28.572	27.305
38	34.522	33.636	32.590	31.404	30.115	28.776	27.436
39	35.215	34.226	33.066	31.762	30.367	28.940	27.537
40	35.880	34.779	33.496	32.073	30.575	29.069	
41	36.513	35.292	33.881	32.340	30.744	29.168	
42	37.113	35.763	34.221	32.563	30.878		
43	37.677	36.190	34.516	32.747	30.982		
44	38.202	36.573	34.767	32.895			
45	38.688	36.913	34.978	33.012			
46	39.133	37.209	35.152	33.102			
47	39.535	37.463	35.292				
48	39.895	37.678	35.403				
49	40.213	37.856					
50	40.489	38.001					

TABLE C-4

	Cycle Ratio, P					
N	.038	.040	.042	.044	.046	.048
1	.963	.962	.960	.958	.956	.954
2	1.924	1.920	1.916	1.912	1.908	1.904
3	2.882	2.876	2.869	2.863	2.857	2.850
4	3.837	3.828	3.819	3.809	3.800	3.791
5	4.788	4.776	4.764	4.751	4.738	4.726
6	5.736	5.720	5.704	5.687	5.671	5.654
7	6.679	6.659	6.639	6.618	6.597	6.575
8	7.618	7.593	7.567	7.541	7.515	7.488
9	8.551	8.520	8.489	8.457	8.424	8.391
10	9.478	9.441	9.403	9.364	9.324	9.283
11	10.399	10.354	10.309	10.261	10.213	10.162
12	11.312	11.259	11.204	11.147	11.088	11.027
13	12.216	12.154	12.088	12.020	11.949	11.876
14	13.111	13.037	12.959	12.878	12.793	12.704
15	13.996	13.908	13.816	13.719	13.618	13.511
16	14.868	14.765	14.656	14.541	14.420	14.293
17	15.726	15.605	15.476	15.340	15.196	15.045
18	16.568	16.426	16.275	16.114	15.944	15.764
19	17.393	17.226	17.049	16.859	16.658	16.445
20	18.197	18.003	17.794	17.572	17.335	17.085
21	18.979	18.752	18.508	18.247	17.970	17.678
22	19.734	19.470	19.186	18.882	18.560	18.220
23	20.461	20.154	19.824	19.472	19.099	18.708
24	21.155	20.800	20.419	20.013	19.585	19.138
25	21.813	21.404	20.966	20.501	20.014	19.511
26	22.431	21.963	21.462	20.935	20.387	19.825
27	23.006	22.472	21.905	21.312	20.702	20.083
28	23.534	22.930	22.293	21.634	20.963	20.290
29	24.012	23.334	22.627	21.902	21.172	20.450
30	24.439	23.685	22.906	22.119	21.336	20.570
31	24.814	23.983	23.135	22.290	21.460	20.658
32	25.136	24.230	23.318	22.421	21.551	
33	25.407	24.430	23.460	22.518		
34	25.631	24.588	23.567			
35	25.811	24.709	23.646			
36	25.952	24.799				
37	26.059					
38	26.139					

TABLE C-5

	Cycle Ratio, P					
N	.050	.052	.054	.056	.058	.060
1	.952	.951	.949	.947	.945	.943
2	1.900	1.897	1.893	1.889	1.885	1.881
3	2.844	2.837	2.831	2.824	2.818	2.811
4	3.781	3.772	3.762	3.753	3.743	3.734
5	4.713	4.700	4.687	4.674	4.660	4.647
6	5.637	5.620	5.603	5.585	5.567	5.549
7	6.553	6.531	6.509	6.486	6.463	6.440
8	7.460	7.432	7.404	7.375	7.345	7.315
9	8.357	8.322	8.286	8.249	8.212	8.174
10	9.241	9.197	9.153	9.107	9.060	9.012
11	10.111	10.057	10.002	9.946	9.888	9.828
12	10.964	10.899	10.832	10.762	10.690	10.616
13	11.799	11.720	11.637	11.552	11.464	11.373
14	12.612	12.516	12.416	12.312	12.205	12.094
15	13.400	13.284	13.163	13.038	12.908	12.773
16	14.159	14.020	13.875	13.723	13.567	13.405
17	14.886	14.719	14.545	14.364	14.177	13.983
18	15.575	15.377	15.170	14.955	14.733	14.504
19	16.222	15.988	15.744	15.491	15.231	14.964
20	16.822	16.548	16.262	15.968	15.666	15.359
21	17.371	17.052	16.722	16.383	16.038	15.690
22	17.865	17.498	17.120	16.736	16.348	15.958
23	18.301	17.883	17.457	17.027	16.596	16.168
24	18.678	18.209	17.734	17.260	16.790	16.328
25	18.996	18.475	17.955	17.441	16.936	16.444
26	19.256	18.688	18.126	17.576	17.041	16.525
27	19.464	18.852	18.253	17.673	17.114	
28	19.624	18.974	18.345	17.740		
29	19.744	19.062				
30	19.831					

TABLE C-6

N	Cycle Ratio, P						
N	.062	.065	.070	.075	.080	.085	.090
1	.942	.939	.935	.930	.926	.922	.917
2	1.877	1.871	1.861	1.851	1.842	1.832	1.822
3	2.805	2.795	2.779	2.762	2.746	2.729	2.713
4	3.724	3.709	3.685	3.660	3.635	3.610	3.585
5	4.633	4.613	4.579	4.543	4.508	4.472	4.435
6	5.531	5.504	5.457	5.409	5.360	5.310	5.259
7	6.416	6.380	6.318	6.254	6.188	6.121	6.052
8	7.285	7.238	7.157	7.074	6.987	6.898	6.807
9	8.135	8.075	7.972	7.864	7.752	7.636	7.516
10	8,963	8.888	8.757	8.619	8.476	8.327	8.173
11	9.767	9.672	9.507	9.333	9.152	8.963	8.769
12	10.541	10.423	10.217	10.000	9.773	9.538	9.297
13	11.280	11.135	10.880	10.612	10.333	10.044	9.750
14	11.980	11.802	11.491	11.164	10.824	10.477	10.127
15	12.634	12.419	12.042	11.648	11.244	10.835	10.427
16	13.238	12.979	12.529	12.063	11.590	11.118	10.655
17	13.784	13.477	12.948	12.406	11.865	11.334	10.821
18	14.270	13.911	13.297	12.680	12.074	11.490	10.935
19	14.692	14.277	13.579	12.890	12.226	11.597	11.009
20	15.048	14.577	13.798	13.044	12.331	11.667	
21	15.339	14.815	13.962	13.152	12.400		
22	15.570	14.996	14.078	13.225			
23	15.746	15.129	14.158				
24	15.876	15.233					
25	15.967	15.286					

N	.095	.100	.105	.110	.115	.120	.130
1	.913	.909	.905	.901	.897	.893	.885
2	1.813	1.003	1.794	1.784	1.775	1.765	1.747
3	2.696	2.679	2.663	2.646	2.630	2.613	2.580
4	3.559	3.533	3.508	3.482	3.455	3.429	3.377
5	4.398	4.360	4.323	4.204	4.246	4.207	4.129
6	5.207	5.155	5.101	5.047	4.993	4.938	4.826
7	5.981	5.910	5.837	5.762	5.687	5.611	5.458
8	6.713	6.617	6.519	6.420	6.319	6.218	6.013
9	7.393	7.268	7.140	7.010	6.879	6.747	6.483
10	8.015	7.854	7.691	7.525	7.359	7.193	6.862
11	8.570	8.368	8.163	7.958	7.753	7.550	7.153
12	9.051	8.803	8.554	8.307	8.063	7.823	7.361
13	9.453	9.157	8.863	8.574	8.292	8.018	7.501
14	9.777	9.432	9.094	8.767	8.452	8.150	7.589
15	10.025	9.635	9.259	8.899	8.557	8.233	7.640
16	10.207	9.777	9.369	8.983	8.621	8.281	
17	10.332	9.871	9.438	9.034			
18	10.414	9.929					
19	10.464						

TABLE C-7

Cycle Ratio, P

N	.140	.150	.160	.170	.180	.190	.200
1	.877	.870	.862	.855	.847	.840	.833
2	1.728	1.710	1.692	1.674	1.656	1.639	1.622
3	2.547	2.514	2.481	2.448	2.416	2.384	2.352
4	3.324	3.271	3.218	3.165	3.112	3.060	3.008
5	4.050	3.970	3.891	3.811	3.732	3.653	3.576
6	4.713	4.600	4.486	4.373	4.261	4.150	4.041
7	5.302	5.147	4.992	4.839	4.688	4.541	4.397
8	5.807	5.602	5.401	5.204	5.012	4.827	4.650
9	6.220	5.962	5.711	5.470	5.239	5.020	4.813
10	6.539	6.228	5.930	5.649	5.385	5.138	4.908
11	6.771	6.411	6.074	5.760	5.471	5.204	4.959
12	6.928	6.527	6.159	5.823	5.517	5.237	
13	7.027	6.596	6.207	5.856			
14	7.084	6.633					

N	.220	.240	.260	.280	.300	.320	.350
1	.820	.806	.794	.781	.769	.758	.741
2	1.588	1.555	1.522	1.491	1.461	1.431	1.388
3	2.289	2.227	2.167	2.109	2.052	1.997	1.918
4	2.906	2.806	2.709	2.615	2.525	2.438	2.314
5	3.423	3.275	3.134	2.998	2.869	2.747	2.577
6	3.829	3.628	3.438	3.260	3.094	2.940	2.730
7	4.124	3.869	3.634	3.419	3.223	3.045	2.806
8	4.318	4.017	3.747	3.505	3.288	3.094	2.839
9	4.433	4.099	3.804	3.545	3.317	3.114	
10	4.495	4.138	3.830				

N	.400	.450	.500	.550	.600	.650	.700
1	.714	.690	.667	.645	.625	.606	.588
2	1.321	1.258	1.200	1.146	1.096	1.049	1.006
3	1.795	1.682	1.579	1.485	1.400	1.323	1.252
4	2.125	1.958	1.810	1.679	1.563	1.460	1.368
5	2.326	2.111	1.927	1.769	1.633	1.515	1.412
6	2.429	2.182	1.976	1.803	1.657	1.532	1.425
7	2.475	2.209	1.993				
8	2.492						

N	.750	.800	.850	.900	.950	.995
1	.571	.556	.541	.526	.513	.501
2	.966	.928	.893	.860	.829	.803
3	1.188	1.129	1.075	1.025	.980	.942
4	1.286	1.213	1.147	1.088	1.034	.989
5	1.321	1.241	1.170	1.106	1.049	1.002

INDEX